AF531708

Extra-Ordinary

Extra-Ordinary

Stories of Science in Everyday Life

Aparna Agarwal

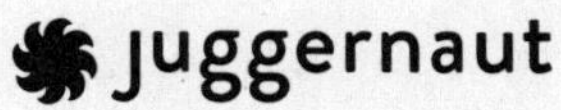

JUGGERNAUT BOOKS
C-I-128, First Floor, Sangam Vihar, Near Holi Chowk,
New Delhi 110080, India

First published by Juggernaut Books 2026

10 9 8 7 6 5 4 3 2 1

P-ISBN: 9789353454395
E-ISBN: 9789353455538

Typeset in Adobe Caslon Pro by R. Ajith Kumar, Noida

Printed at Thomson Press India Private Limited

To my Mamaji, the late Mr Vishwanathan Nair.
Hope this makes you smile :)

Contents

Prologue

Unusual stories. Science. Magic and the mundane. How do we humans see the world and why do we put those two things – magic and science – in different buckets? What connects them? What do we learn and how do we feel when we change an artist's easel for the scientist's microscope?

Hello, dear reader. You don't know me, or maybe you do, but let's assume you don't (I do want strangers to read this book after all!).

This book is about some of the things that I find both beautiful in the scientific context and utterly magical when I wear my amateur artist's cap. Many of these are ordinary things, but indulge me as we walk into a realm where science and magic mix.

Actually, strike that, because it's not just the realm of imagination where these domains overlap. As the science-fiction writer and satellite-communication pioneer Arthur C. Clarke once said, 'Any sufficiently advanced technology is indistinguishable from magic.'

It's a reminder that magic and science aren't always as far apart in our reality as we think. Sometimes, all it takes is a shift in perspective for the ordinary to start feeling like magic.

For example, it would seem impossibly magical to somebody from a society without a written language if they saw us communicate across time and space by simply using these strange symbols we call letters. And yet, to us, that's ordinary, and therefore somehow, not magical.

Why isn't that magic? Is it any less magical just because we know how it works? I don't think so. To explain how I feel, let me share a quote from a famous scientist, Richard Feynman:

> I have a friend who's an artist and [he] has sometimes taken a view which I don't agree with very well. He'll hold up a flower and say, 'look how beautiful it is,' and I'll agree. Then he says, 'I as an artist can see how beautiful this is but you as a scientist take this all apart and it becomes a dull thing,' and I think that he's kind of nutty. First of all, the beauty that he sees is available to other people and to me too, I believe. Although I may not be quite as refined aesthetically as he is ... I can appreciate the beauty of a flower. At the same time, I see much more about the flower than he sees. I could imagine the cells in there, the complicated actions inside, which also have a beauty. I mean it's not just beauty at this dimension, at one centimetre; there's also beauty at smaller dimensions, the inner structure, also the processes. The fact that the colours in the flower evolved in order to attract insects to pollinate it is interesting; it means that insects can see the colour. It adds a question: does this aesthetic sense also exist in the lower forms? Why is it aesthetic? All kinds of interesting questions which science knowledge

only adds to the excitement, the mystery and the awe of a flower. It only adds. I don't understand how it subtracts.

This is exactly how I feel, and it's one of the themes that runs through the book. I find the everyday world around me fascinating as I look at the magic hidden behind everything we dismiss as ordinary.

Since my world revolves around my daily routine, and my daily routine revolves around my dog Oreo, he is an integral part of these stories. Indeed, these stories all arose organically from the things Oreo and I do on Sundays and I've taken a little artistic licence in writing as though they all happened on a single ordinary yet magical Sunday.

It's time for you, dear reader, to meet the dashing Oreo Pupperwal, first of his name. He briefly considered being called Sir Oreo Pupperwal, but ultimately decided he didn't need the honorific. He self identifies as 'most important', and it has worked out great for him so far. Case in point, he is the only one who can boss around the strong, independent woman that I pretend to be.

Oreo is a four-year-old black-and-white dog who has the aloof yet demanding personality of a black cat – complete with judgmental stares and selective hearing. Yet, he howls like a husky, watches me like a sheepdog and has more than once been mistaken for an odd-shaped Dalmatian because of his spots. In short, he's a lot like his human – a mix of many things, yet totally unique.

We've been together since 2020, with him bossing his way into my heart from our very first encounter. This tiny puppy ran to me and hid behind my legs, not because he wanted my attention but because I made a convenient human shield against other humans, two big dogs and one very judgy cat. I admired his priorities. And was secretly overjoyed with the quiet honour of being chosen as the safe human.

I sample much of the world around us in his company and share in his delight as we go through our daily routine of cuddles, walks and play. He reminds me to literally stop and smell the flowers, and I tell him about the hidden science behind things. I tell him things I've found hidden in obscure corners of the Internet and buried in books. And he listens to all of this patiently, even if it sounds like alien gibberish to him.

Sometimes I do feel like an alien with two hearts, one devoted to being an artist and the other delighted by dabbling in the sciences. I often find myself gravitating towards the stories that glue these two together.

So, I would love to share with you stories from the world of science, which, to me, are also stories from the world of art. Stories that make me love the fact that we exist on our little blue dot at a time when we can – to quote yet another great scientist – stand on the shoulders of giants and look at our world from a completely unique perspective. In my case, that also often involves looking at the world (and smelling it!) from a canine perspective.

I can't wait to share these stories with you. Stories that are part science, part daydreams, guided by curiosity, wonder and, occasionally, the wise instincts of a four-legged friend who knows that some of the most profound discoveries begin with simply stopping to notice the mundane.

1

The Immortal Eternal Apple

In the beginning, there was an apple tree. It had been standing there for centuries, and it was the first of its kind, and also the last of its kind. Or was it? The details are a bit fuzzy, so let's rewind!

In the beginning, there was the calm, cool morning of a peaceful day of rest. It was Sunday, the day when everyone's allowed to sleep in a little bit longer.

This, of course, is true only when you don't have a boss that insists you wake up early on a Sunday. In my case, that boss is a dog who doesn't get the concept of Sundays. So, seven days a week, my alarm clock is a well-loved and rather chewed up ball being pushed into my face to remind me that it is time for my dog Oreo's breakfast.

As I walk into the kitchen to prepare his breakfast, I am reminded of the fact that I have to make my own breakfast as well. I sigh, vaguely remembering that as a child I couldn't wait to grow up and eat whatever I want, whenever I want. I glance around, dreaming of something deliciously indulgent, but alas, my post-workout grocery shopping meant my choices are limited to only 'healthy' options. My gaze wanders until it lands on something red – an apple. Now, as an adult I am burdened with the knowledge that all that freedom also means sometimes eating an apple.

As I hold this apple in my hand, my brain offers up a proverb, because of course.

It's the one that goes 'an apple a day keeps the doctor away'. This, in turn, leads to its own spiral with a series of interesting metaphysical questions. Given that I am a doctor (at least technically), am I currently being kept away from myself because of these apples? How does that even work? Am I just using this as an excuse to order something instead of eating the fruit in front of me? Ah. Questions.

As I stare at the apple, wondering whether I should cut it, grate it or eat it whole, I let myself slide further into thought. Whenever I look at any object, my brain has the habit of showing me all the interesting titbits it has collected over the years on the topic, like an eager friend who's been asked about something they know far too much about and simply can't stop talking about.

Come, let's talk about apples.

Apples are fascinating, delicious and quite abundant. There are so many varieties of them – there are the really small Kashmiri apples, the almost purple red Kinnaur apples, the bright red royal apples and even the almost yellow golden apples. These are just the varieties grown in Himachal Pradesh. If we look across the world, we have the deliciously sour Granny Smith apples, the extremely sweet honeycrisps or the pink lady apples that to me evoke the image of a cartoon woman in a pink ballgown handing out candied apples.

Apples are a huge part of our collective imagination, from the poisoned apple of Snow White to the Biblical apple of Eden from the Tree of Knowledge,[1] or the famed apple falling on Newton's head, causing him to discover gravity.[2]

For me personally, apples evoke memories of my childhood car rides through Himachal, where orchards lined the road – I can smell them even now. Whenever I remember walking in an apple orchard with trees laden with fruit, I understand why so many poems, stories and myths feature the fruit.

To delve further, let's enter the land of imagination. Here, the unexpected often happens, allowing us to see things we might not see in our everyday lives.

Come, let's walk through an apple orchard. I walk up to the nearest tree and wonder, 'Does this tree have a name?'

The tree turns to me and says, 'Hello. I'm Komal.' I'm startled until I remember that this is the land of imagination after all, where trees can talk. So I say hello to Komal the Tree, and walk over to the next tree.

'Hello, what's your name?' I ask.

---

1 Sadly for the apple, it is very unfortunately associated today with that story, when in reality, the word 'apple' just meant any fruit in Middle English. Up till the seventeenth century, 'apple' was a generic term for all fruit apart from berries, and even included nuts. Dates were finger apples, and bananas were called 'ye appel of paradis'. This also shows the influence art can have on collective memory, as most of the early illustrations of the Garden of Eden depict an apple, burning that image into the collective psyche.

2 Of course, this is just a story. Newton did not discover gravity just because an apple fell on his head. Yet it is a story we like telling and retelling, because it sounds so good.

The tree looks at me and replies, 'Hello. I'm Komal!'

Bewildered, I ask another tree, and it also says, 'Hello. I'm Komal.'

Confused?

Well, so would I be – if I didn't know that these are apple trees. Which means that all of these trees are, in fact, clones!

So, what does it mean that all these trees are clones? Is this a sci-fi story set in the distant future where we have for some reason cloned … apples?

Not really! It is a story about clones, but it's set very much in our time and, indeed, part of our history as a species that evolved from being hunter-gatherers and became farmers. These trees – and in fact all trees that bear a certain variety of apple – are clones of a single ancestral tree that initially grew that kind of apple. The same tree has been grown and regrown over and over again for centuries, clones producing the exact same fruit!

Indeed, in the beginning, there was an apple tree, and it seems it is still here!

This brings us to even more questions. Why are all these trees clones? What does it mean when we say a plant is a clone? Can we not just grow apples out of seeds? As I think of all these things, I discover I'm not actually eating the apple in front of me.

Well, since I'm not eating it anyway, let me fully commit. Come, sit with me under the shade of this lovely apple tree of our imagination as I tell you all about apples, and much more. This is a story about how humans should not have been able to domesticate apples, and how – spoiler alert – we did it anyway.

I know we were talking about clones and apples, so how is it a story about domestication? And what even is domestication? It rings a bell as it is a moderately well-known word – many people may have heard of the concept, but not everyone is familiar with how it actually works, so let's talk about it.

'Domestication' is a complex word; when I first read it as a tiny human, I couldn't pronounce it correctly. However, my mother taught me a trick to tackle such big words instead of being scared of them. She told me, 'If you encounter a big word that you don't know the meaning of, or are intimidated by, just break it down, sound it out, and if you still don't understand it, use the dictionary or ask someone who knows how to pronounce it. Slowly but steadily, you will learn so much more about it.'

Well, this trick works with concepts, ideas and scientific details as well as words. It's all about breaking something we don't understand down into chunks that we can understand and looking for things we don't know.

So, let's break it down.

Dom-est-ica-tion comes from the mediaeval Latin word *domesticare*, which loosely translates to 'to tame' or 'to dwell in the house'. When it comes to animals like dogs – yes! – domestication refers to them living in close quarters with humans, while with plants, it often refers to our ability to grow them in large quantities via agriculture.

Let us talk about domestication in plants. The way domestication in plants usually works is simple. We take a tree that grows the fruit we like and collect the seeds from the tastiest fruits. We then plant these seeds to grow the next generation of trees. If all goes well, in a few generations, we get trees that mostly grow the delicious fruits that we like.

That, however, is a very simplified view of what really happens.

In reality, very few things are truly that simple. Whether we get delicious fruits from the seeds of delicious fruits depends on several things. One of them is heritability. Heritability is a measure of the chance that any desirable feature that you want – also known as a 'trait' – gets passed along to the next generation from the parent.

Let's say that every time you plant a seed from a delicious fruit, you always get a tree that also grows those delicious fruits. That means this trait that we like, has a 100 per cent heritability. However, if the chances of getting a delicious fruit from this seed are lower – say, only one in 10 trees grows such fruits – then the heritability is down to 10 per cent. As the chances get lower and lower, the uncertainty increases manifold. Imagine being a farmer, trying to grow fruit to sell, and never knowing whether the tree you planted will yield edible fruit or not.

All this is not to say that it is not possible to domesticate plants. That, in essence, is what we have done for many crops that we farm. If you sow wheat or barley seeds, you are guaranteed to get wheat or barley that is very similar to the seeds that were planted, almost 100 per cent of the time.

Which brings us to the golden question: What are the chances of getting a golden apple if we plant a golden apple seed?[3] The answer, actually, is virtually none. Apple seeds have a really low degree of inherent heritability. This means that the chances of getting similar traits to the parents are overall lower. A tree grown from the seed of an apple can yield edible but very different fruits, or fruits that taste absolutely horrible. If your livelihood depends on getting tasty fruits, apple seeds are probably the worst of the lot to bet on.

Now comes the second problem. 'Deliciousness' is not really a single trait. It is a combination of traits.

Why does that matter? Well, let me give you an example from our everyday life.

Every time we follow a recipe, it involves adding several different ingredients. All of these ingredients need to be added in a specific ratio and often in a specific order. In addition to that, the ingredients need to be chopped a certain way, cooked for a specific amount of time, mixed with specific spices and served hot or cold. All of these factors affect the taste of the final product.

To take a simple example, a dosa is definitely not the same as a dish made of rice, dals and other ingredients put together anyhow. In fact, even using the same batter, the dosa will be crispy when the batter is spread thin, fluffy when poured as a set dosa and taste completely different based on the amount

[3] Three 'goldens' in one sentence deserves some bonus trivia, so here you go. In Greek mythology, the Apple of Discord was a golden apple thrown by Eris, the goddess of strife, during the wedding of Peleus and Thetis. This act sparked a vanity-fuelled rivalry among the goddesses Hera, Athena and Aphrodite, which eventually culminated in the Trojan War.

of butter or ghee used. So, the 'taste' of the final product is greater than the sum of its parts, with every individual element having a considerable influence on the final product.

Similarly, for the apple to be delicious, we need a combination of sweetness, texture, smell, shape and several other factors. This means that for apples from a tree to pass our criteria of 'deliciousness', they have to inherit all those traits at a relatively high rate, and the lack of any single one of these traits may lead to an apple that doesn't quite make the cut.

Since we already know that apples have a low heritability for most traits, and you add the number of traits required for deliciousness, it becomes apparent that growing an edible apple out of a seed is a really big challenge.

Moreover, apple flowers do not self-pollinate. For any fruit to grow out of a flower, the flower needs to be fertilized by combining the male gamete (known as pollen) with the female gamete (known as the ovule). This process is called pollination.

For apples, the plants need an external source of pollen, which usually comes from nearby crab apple trees, or from another apple variety planted nearby. This means that while the fruit looks like the parent tree, the DNA inside is from two different sources. This process further increases the inherent variability that we find in apples in every generation, thus leading to the fruits being dissimilar to the parent.

So what would happen if we took a lot of seeds and kept growing them until we found seeds that grow trees with the traits we want to breed for? Well, a group of German scientists did this experiment, starting with 52,000 seeds from different types of apples, trying to grow apple seedlings that were scab resistant. After 26 years of growing seedlings, screening them in the lab and then grafting them onto rootstocks and evaluating them in the field, do you want to guess how many varieties of apple fit their criteria? Three! Out of these, how many healthy cultivars were suitable for cultivation? Just one! Like I said earlier, growing apples from seeds isn't a gamble for the feeble-hearted.

So, as I mentioned, we should not have been able to domesticate apples.

But we did. How did we manage it? You could say that humans that domesticated apples did not know that they shouldn't have been able to grow them, just like the bumblebee that didn't know that it shouldn't be able to fly because of aerodynamics. They imagined a universe where they could harvest apples and made that a reality.

Indeed, many inventions have resulted from the fact that humans are very persistent. It is this same persistence that fuels all of science, with us looking for ways to travel to the stars, and also to look at the tiny particles within an atom. If we want something, as a species we keep trying to find ways to get it. So how did we manage to get crops of apples? And why are they 'clones'?

It's because we used grafting, or something called 'clonal propagation'. This is the process that made all the Komals clones.

Grafting is like something out of a sci-fi movie – if we didn't already know it was a thing, we'd probably assume aliens came up with it. Picture this: From a branch from one plant, you can get a brand new tree that can grow leaves, flowers and even fruits! If humans could pull off something like that, you'd be walking down the street, see a stray arm or leg growing into a full-on person, and just say, 'Oh, cool, they're grafting.' Sounds bizarre, right? Well, plants have a secret superpower that makes it all possible: pluripotency.

Certain – but not all – plant cells have the ability to become almost any type of cell the plant needs. In grafting, it is this ability that is harnessed. A part of the desired plant we want – often the stem – is joined together to the root system of another tree. This part is vital, and it is where farmers with a lot of skill combine the cambium – the part of the plant that rapidly divides – of the two trees together by creating complementary cuts and interlacing them. As time passes, the site of the cut starts healing, creating a seamless connection. Before we know it, we have a whole new tree growing from the small section of the stem, which retains all the features of its parent plant, while being nourished by the roots of the plant it was grafted on.

A tree grown like this is a clone of its parent tree. This process bypasses most of the problems of heritability I mentioned above. As long as you pollinate it with the same kind of trees that the original was pollinated with, you are guaranteed a fruit that is very similar to the parent. In addition, grafts usually grow much faster than a tree grown from the seed stage, and this gives farmers the ability to speed through the process of getting a fruit-bearing tree. Now that's something I would like, if my livelihood depended on selling delicious fruit.

So that's how we did it. We took a plant that we shouldn't have been able to domesticate, and we did it anyway. Often when looking at impossible problems, we look at nature for guidance because if we look long enough, we find that an apparently impossible thing like the one we're interested in already exists, and then we can start asking 'how'. There are theories that suggest that early grafting may have been inspired by the extremely rare natural grafts that occur in some plants.

By now, grafting has become quite a commonplace procedure. It's how we grow roses, plums and cherries. But here's where it gets even more mind-blowing – people have mastered grafting to the point where they can grow different varieties of fruit on the *same* tree. By painstakingly grafting multiple varieties on the same tree over many years, people have been able to create Frankenstein-esque trees that have over 250 different varieties of apples! If you want a mental workout, just scour the Internet for an explanation of '250 varieties of apples on a single tree'.

In fact, grafting – also known as vegetative propagation – also known as clonal propagation – is something that humans have been using for a long time around the world, from monks in Europe to farmers in Asia. It's used to cultivate many types of plants, not just trees with low heritability like apples, which we can't easily breed for the characteristics we want. By some estimates, over 75 per cent of all types of perennial[4] fruits that humans grow commercially involve some form of grafting. Talk about hacking nature!

---

[4] 'Perennial' comes from the Latin perennis, which means 'lasting through the year'. For plants, it refers to any plants that grow for longer than two years. Fruit trees like apples, once planted, can last for several years and are thus called perennials, as opposed to plants like wheat which need to be grown every year and harvested, thus called annuals.

It is also how we grow seedless varieties of plants like the Cavendish banana – named after William George Spencer Cavendish. Just like apples, Cavendish bananas are all grown from clones of a single seedless plant variety, first grown in Britain in 1834. The same is true for seedless navel oranges and many wine-producing varieties of grapes. So in this specific case, you can actually compare apples to oranges, or even grapes!

Of course, as with any plant, environmental factors like soil conditions, humidity and water source can have a huge impact on the taste of the fruit. Think of it like this: You may be following the exact same recipe as one you've made before, but if the quality of individual ingredients is poor, the resulting food will still taste very different. It is similar in plants, as they use the nutrients from the soil to create the fruit every fruiting season. This means that everything else being the same, there will still be a difference in how the fruits taste every year if the environmental conditions change. This is also why certain wines from specific years cost a lot more than others, as the inherent quality of the grapes changes from year to year, due to minute differences in soil, weather and other environmental conditions. (There are some who believe that wine pricing is all an elaborate scam and even experts can't really tell the difference, but that is a different subject for another time.)

Grafting is also the method that allows us to bring back ghosts! I don't mean bringing back the dead, but something scientists refer to as ghosts. More specifically, fruits like the avocado are called 'ghosts of the evolutionary past'. That's because the natural process through which avocado seeds turned into trees involved some really large mammals like the mammoth or the mastodon. These massive creatures would eat the fruit whole, and the large avocado seeds would pass through their digestive systems, only to be deposited miles away in their droppings – a perfect way for the tree to spread and germinate.

Unfortunately for avocados, those giant mammals disappeared about 10,000 years ago – so, you'd think the avocado would have gone extinct, too. After all, its huge seeds weren't suited for smaller animals to spread. Yet the avocado survived, which brings us to the following question: How does a plant survive if it has a seed that requires it be passed through the guts of the extinct mastodon to germinate?

Well, one reason it survives is because avocado trees naturally have a long lifespan, which allowed them to survive for a very long time even though the natural dispersers of their fruit were no longer around. Even if no new trees were growing, the older ones that remained kept growing and quietly producing fruit year after year. In modern times, it has survived because of humans who painstakingly grew it. Unlike the apple, we didn't initially grow avocado trees using grafting, but from seeds – a tricky, highly variable process that likely took a lot of trial, error and patience. To mimic what once happened in the guts of those extinct animals, people had to soften or remove the outer seed coat, and sometimes even score or pierce the seeds to help them germinate. Even today, if you Google how to grow an avocado plant, the first step would be to pierce the seed with toothpicks or score it with a knife.

Why did humans do it? The answer is unclear. Maybe

because it is an excellent source of healthy fats. Maybe because it became a part of the diets and cultural norms of several Central American civilizations. The ancient Aztecs, for example, apparently considered the fruit an aphrodisiac and something that boosted fertility.[5] Or maybe – just maybe – humans really like eating guacamole, and thus learnt to grow the fruit that gives it its characteristic taste. Whatever the reason, the truth remains: The avocado exists today because humans decided it was too delicious to lose.

Here's where grafting enters the picture though. Today, most commercial avocado varieties, especially the popular Hass avocado, are propagated through grafting. This ensures that each tree produces fruit with the same taste, texture and quality. Every Hass avocado you eat essentially comes from clones of a tree that was discovered in the 1920s by a California postman named Rudolph Hass. By grafting branches from this original tree, growers have been preserving the ghosts of the avocado past for future generations to enjoy.

So, the next time you take a bite of an apple, I hope you smile at the thought of how every Granny Smith apple comes from clones of a tree found in 1868, growing as a chance seedling in a compost pile in the orchard of farmer Maria Ann 'Granny' Smith in Ryde, New South Wales. I also hope that it makes you marvel at the sheer power of the human mind and

---

[5] The word 'avocado' comes from the Spanish aguacate, which derives from the Nahuatl (Mexican) word āhuacatl, both of which anecdotally refer to a certain part of a bull's anatomy (yes, that part!) that the fruits resemble. No wonder, then, that the ancient Aztecs considered these fruits a symbol of fertility.

our imagination – our striking ability to tell ourselves stories in which we can imagine the impossible being possible and then figure out a way to do it in real life.

For instance, if you thought growing plants without seeds was impressive, wait till I tell you that we figured out how to grow plants without any soil, and from a very small number of cells. This incredible feat is achieved through a technique known as tissue culture. In this process, we can take just a few cells from a plant – sometimes as small as a microscopic piece of tissue – and use them to grow part of a, or even an entire, new plant. What makes this possible is that very same power of pluripotency. To remind you, pluripotency refers to the ability that some plant cells have the potential to regenerate into an entire plant, given the right conditions. This is why tissue culture works – it taps into the natural flexibility of plant cells to grow into whatever is needed.

This ability is a game changer for agriculture, conservation and even plant research. Through tissue culture, we can create clones of plants with desirable traits like disease resistance or improved growth rates, or rescue endangered species that are difficult to propagate naturally. It allows us to mass-produce plants that are genetically identical, ensuring consistency in crops or helping preserve rare species that are threatened in the wild. Anecdotally, one of the key factors that led to the development of the art and science of tissue culture was a desperate need during World War II. The British and Japanese nurses who were treating soldiers in the war apparently used coconut water for infusions, as sterile saline – which is what is typically used – was in very short supply. Since coconut water is naturally sterile and contains most of the vitamins and nutrients required for cells to grow, it was a useful alternative in the time of war. That said, it was only used in times of emergency, and then too as a very temporary alternative.

However, the story goes that this led scientists who were trying to grow plant cells to use coconut water as a medium, and that actually ended up working much better than any existing mediums![6] In fact, coconut water was used as the base for tissue culture for a long time, until artificial nutrient mediums were developed.

Coconut water is rich in vitamins, amino acids and sugars, making it an excellent natural medium for supporting plant cell growth and for rehydrating yourself on a warm day. While synthetic nutrient solutions have become more common, coconut water remains a valuable tool in certain applications due to its natural composition.

So far, I've told you all about the marvellous ingenuity of humans that has allowed us to do what once seemed almost impossible. But it's important to acknowledge that not everything is sunshine and rainbows. There are significant downsides to growing plants through grafting and cloning, particularly when it comes to reducing or even eliminating their natural diversity.

While apples in nature boast an impressive genetic diversity, apple trees grown as cultivars[7] often suffer from very

---

[6] Again, while both parts of this story are independently true, we do not have any concrete proof that shows that one led to the other. Yet, like with Newton and the apple, this is another story that is both compelling and fun so we love to propagate it.

[7] Cultivar is just a fancy word for saying 'grown as a specific variety of apple'.

low genetic diversity. This lack of diversity can make them particularly vulnerable to a host of diseases and pests that could otherwise be mitigated by nature's built-in safeguards. It is akin to having the exact same password across many different websites. Once a hacker has access to one, they can wipe out your entire digital existence, since the password gives them access to everything else.

In the wild, the genetic diversity among different apple trees means that when a disease strikes, not all trees are affected equally. Some might show resistance while others succumb quickly, creating a balance that helps maintain the overall health of the ecosystem. However, in the world of clones, where genetically identical plants are grown side by side, a disease can sweep through an orchard like wildfire. Since all the trees share the same genetic makeup, any disease that targets one tree will likely affect all the others in the same way. This can lead to catastrophic losses in a single growing season, with entire orchards wiped out by a single disease outbreak. The consequences can be devastating, not just for farmers but for food supply as well.

This isn't just the case for apples. As I mentioned before, the Cavendish banana has a global presence in supermarkets, largely because of its ability to withstand shipping and its consistent taste. However, it being a clone means that it's genetically identical across the globe, leaving it as vulnerable as a lone tree in an orchard full of identical siblings.

A devastating fungal disease known as Tropical Race 4 (TR4) is now threatening Cavendish bananas, and if it spreads unchecked, we could face a banana apocalypse! What's more, this fungal disease is known to be resistant to most known fungicides, which makes its effects all the more dreadful. Something like this can affect the global banana trade, which supports millions of farmers worldwide.[8] This has happened before, with bananas being devastated in the 1950s by a fungus that caused 'Panama disease', which is why the Cavendish, which was Panama-resistant, replaced the Gros Michel, which was not. Moreover, because most cloned plants are genetically identical, they can be particularly susceptible to diseases and pests, leading to the widespread use of fungicides and pesticides. These chemicals may help control outbreaks in the short term, but they can harm beneficial organisms in the soil, disrupt local ecosystems and even lead to pesticide-resistant pests, creating a cycle of dependency that can be tough to break. The environmental impact can be significant, especially when runoff from treated fields contaminates local waterways, affecting wildlife and drinking water quality.

Finally, the pressure to produce uniform crops can push farmers to prioritize short-term yields over long-term sustainability. In their quest for the perfect apple or the ideal banana, they might overlook the importance of maintaining a diverse range of plants that can adapt to changing conditions. While grafting and cloning can yield impressive results, over-reliance on these techniques can undermine the very resilience we need to navigate the challenges of climate change and biodiversity loss.

But wait! There's hope on the horizon! As we face these challenges, communities all over the world are beginning to revive wild varieties and harness their incredible genetic

[8] Not to mention the fictional but very important species of Minions!

diversity. For instance, in India, initiatives are underway to bring back traditional fruit varieties that have long been neglected. Take the blackberry, or jamun, for example. Once a common sight in Indian orchards, its wild cousins are being cultivated again, not just for their unique flavour but also for their nutritional benefits, like high antioxidant content. If you have been to the south of India, you may also have eaten the traditional yelakki (also known as elakki or elaichi) bananas, known for their characteristic sweet taste.

By reintroducing these wild varieties – which we once ate, but discarded for easier-to-produce cloned crops – into farming systems, we're not only preserving biodiversity but also enhancing the resilience of our crops. The problem of over-reliance on pesticides and growing crops that severely lack genetic diversity is not limited to fruit trees. Many modern-day high-yield varieties of grains also lack diversity and rely heavily on pesticides for their yield. Slowly but surely, though, people are rediscovering alternatives by looking at the past. Farmers are rediscovering the joys of planting traditional varieties of rice like swarna or sona masoori in the south, the low glycemic index boasting kala namak rice in Uttar Pradesh or the many varieties of black rice found in Northeast India, which have adapted over generations to local conditions. These grains are not only delicious but also carry a wealth of genetic traits that can help them withstand pests and climate fluctuations as well as provide nutritional benefits.

India is also seeing a revival of ancient pulses and legumes that were once staples in local diets. Varieties like kharif urad and mung beans are being promoted by agricultural universities and NGOs, which recognize their ability to enrich the soil and contribute to sustainable farming practices. Millets are being recognized as nutritional superheroes around the world, with India leading the charge. These wild varieties, once pushed aside in favour of high-yield crops, are being embraced for their unique flavours, nutritional value and resilience, helping farmers adapt to changing climatic conditions while ensuring food security.

Moreover, community seed banks are springing up across the country. These are places where farmers can save and exchange seeds of traditional and wild varieties. This kind of community exchange increases social bonding, allows peer learning and helps farmers mitigate the risk of planting crops that are highly susceptible to the increasing challenges thrown at them in the face of climate change. This grassroots movement is a game changer! It empowers local communities to reclaim their agricultural heritage and ensures that genetic diversity remains in our fields, not to mention giving our taste buds a more diverse workout. By encouraging farmers to grow a mix of old and new varieties, we're working towards a future where agriculture is not just efficient but also rich in diversity.

So here I am, with even more questions than I started with, and that is often the case with science and with life. What are the long-term implications of these choices we've made? What happens when a particular grafted plant variety becomes the only option available? Will we still have the ability to explore the vast diversity that nature offers? Relying on a narrow range of cultivated plants can create a precarious situation, not just for growers but for ecosystems worldwide.

In essence, while grafting and cloning showcase the incredible capabilities of human ingenuity, they also serve as a reminder of the importance of preserving the genetic diversity that supports healthy ecosystems and resilient food systems. Balancing our desire for uniformity and efficiency with an understanding of the need for diversity is crucial for a sustainable future. So, as I finally cease my meditations on the apple and start munching the delicious one that's lying on my plate, I wonder, if I could go back in time and change things, would I try to change the trajectory that led us to clone the apple trees? I don't have an answer. What about you? What would you have done?

Rather than coming to a decision, for now I shall quote Robert Frost:

> For I have had too much
> Of apple-picking: I am overtired
> Of the great harvest I myself desired.

And with that, I shall exit my imaginary land and return to the present to continue to feed myself, and my thoroughly unimpressed dog who is still waiting for his breakfast.

2

Look Ma, No Blue!

While I finish my morning coffee, Oreo fixes me with an incredulous stare. He has long since finished his breakfast, and is appalled at my inability to keep pace with him as he gobbles down his food.

'What do you want, Oreo?'

He rolls his eyes, and points to his leash. I can almost hear him thinking, 'This human takes forever to train, and she keeps forgetting every day again.' When I don't move, he walks up to me and puts his paw on me and stares intently. I have been given the final warning. I understand when I am defeated. We must go now. So let's do that.

As I put my hands on the leash, the little boy in front of me transforms. His tail wags with an intensity that puts the best propellers to shame, and he jumps up and down like a spring toy wound up to the maximum possible limit. I am in awe of the joy that this boy has every single time we go for a walk.

Living with Oreo has taught me a great deal, and one of the biggest lessons has been learning how to enjoy small moments like this one. As we walk outside, I see remnants of the rain from last night. There are tiny little droplets stuck to the surface of leaves, making everything smell fresh and clean.

Something about this morning brings up a memory, like a sleeper agent being triggered by a keyword. There was a classic drawing I was taught to make as a child – as most Indian kids of my generation probably were – two triangles for mountains, a semi-circular sun between them and a very blue river. I would use brown sketch pens for the mountains, an orange one for the sun and a blue one for the river. The colours were intense, and I used a scale to get the mountains exactly right.

There were several things wrong with that drawing – or rather, the image was somewhat distant from reality. Mountains aren't solid brown, and rivers aren't solid blue with dark blue squiggly lines, are they? In fact, if you were born a few centuries ago, *nothing* was blue.

'That can't be right,' you might say, and Oreo seems to agree, giving me that sceptical side-eye he reserves for when I'm behaving particularly bizarrely (or delaying going for a walk). 'The sky has always been blue, so have the oceans, and surely there must be lots of blue things in nature?' I mean, it's practically a universal example of irrefutable truth.

As if on command, an actress on a loud TV playing in a neighbour's house casually asks, 'Why do we like makeup montages? I don't know. Why is the sky blue?'

My answer to the first question: It gives me joy to see transformations and to understand how colours ever so subtly change the way we look at things.

My answer to the second question? It's pretty much the same as the first one. However, since Oreo has decided that we're taking the scenic route today, let me first tell you why blue didn't exist, and then why that statement is both completely true and blatantly false at the same time. Oh, and also start this walk.

'Sheesh, Oreo, stop pulling on that leash!'

As I look up at the brilliant blue Bangalore[1] sky, I come back to our question. How did the ancient world not have blue when the sky is literally right above their heads?

Before I talk about that, though, let me clarify a linguistic tick which may cause some confusion.

What is colour?

As if reading my thoughts, Oreo suddenly jerks to a stop. Actually, he has stopped to smell a particularly interesting plant, perhaps using his nose to sniff out a message from another dog.

In any case, since we've stopped, let me take this time to answer the question: What is colour?

We use the word 'colour' to mean two different things. One meaning refers to the parts of the spectrum of light that the apes known as human beings can see, and the other refers to compounds that seem to be roughly in the hue of the light spectrum that we have names for.

You see, light is what we call an electromagnetic wave.[2] There are many types of electromagnetic waves, and actually, they are all the same, yet all very different. From radio waves to microwaves, electromagnetic waves are all parts of the full

---

1 Triple alliteration equals bonus trivia – the blue veins in our arms appear to be blue due to something called subsurface scattering.

2 It is also a particle; talk about duality!

spectrum of electromagnetic radiation. A small portion of this is what we refer to as the 'visible spectrum' – the part of the spectrum that is visible to us. Other animals can see parts of the spectrum that aren't visible to us, and others still are limited to seeing only a portion of the spectrum visible to humans. Some animals can 'feel' parts of the spectrum as well as heat.

Now, the visible spectrum is divided into seven colours,[3] starting from violet on one end and going up to red on the other. Ultraviolet rays – aka UV rays[4] – correspond to the wavelength beyond violet, while infrared rays correspond to the wavelength beyond red (and are also the ones responsible for heat).

When we are speaking in terms of this definition of 'colour', most of the colours we see around us are actually the parts of light that objects refuse to absorb. An apple looks red because it's actually absorbing all the other colours and only reflecting red back at us. It's like the apple is saying, 'Nah, you can have the red light, I'll keep the rest.' White things are objects that reflect all wavelengths back. Black objects are the opposite – they're absorbing almost everything.

This is why we are often told to wear white in summers, as the colour helps reflect infrared rays, thus literally pushing

---

[3] The reason why we identify seven colours in the visible spectrum is Newton. He was a devout Christian and did not care for the visible spectrum having only six colours – six being a number often associated with the devil – so he added indigo in order to divide the visible spectrum into seven neat, religiously acceptable colours.

[4] Aka the part of the spectrum that dermatologists hate with a vengeance.

away heat. The opposite is true for black objects – they absorb heat, making them perfect for when you want to sweat that little extra bit. Just make sure it's not too black. Especially not the blackest black.

What is the blackest black, you ask? Guess a digression is in order. Well, in 2014, a company called Surrey NanoSystems released a pigment called 'Vantablack', unofficially known as the 'blackest black', as it absorbed up to 99.965 per cent of visible light. While it has many applications, it is also extremely hard to work with, as any object painted with it becomes extremely hot very quickly.

Vantablack is no longer the blackest black. A new pigment made from carbon nanotubes created at the Massachusetts Institute of Technology now holds that honour, absorbing 99.995 per cent of all light. There's also Black 2.0, which was created as a direct response to Vantablack.

However, there is a story worth sharing about Vantablack.

This a story about Vantablack features some colourful characters. A very well-known artist named Anish Kapoor (the same one who created the infamous bean sculpture in Chicago) struck a deal with the company to let him exclusively use the colour for his artworks only. This sparked a huge controversy in the art world. The idea of restricting a pigment to be used by only one artist seemed outright criminal to many in the art community. In response, another artist, Stuart Semple, created a colour called the 'pinkest pink', with a legal rider: it would be available for use to everyone except Anish Kapoor. Semple later went on to create Black 2.0, which comes with the same legal rider.

So I guess in this case, usable 'colours' for Anish Kapoor

mean everything but the pinkest pink and Black 2.0. Which reminds me, we need to go back to the meaning of colours. Oreo tugs at the leash, clearly wondering why we've stopped for my impromptu physics lesson. 'Just a minute,' I tell him, 'this gets interesting.'

Going back to the two meanings of colour – as I've mentioned, colour is a result of the wavelength of light that an object reflects. How do objects do this though? How do they absorb some wavelengths of light while reflecting others? This is because of pigments. Pigments are molecules or complexes of molecules that absorb a certain portion of the spectrum, while reflecting others.

This is the other meaning of the word 'colour' that we use. The primary colours we were taught in art class (red, green and yellow) do not refer to the wavelengths but these pigments. If you combine all three of the primary colours' wavelengths, you get white. If you do the same with colour pigments, you get a muddy grey or brownish colour depending on the base chemicals used in the pigments.

Now that we've taken a detour to talk about what colour is, let's come back to my earlier claim: 'Blue didn't exist!'

Now, while that's only a three-word sentence, it also, like most things discussed in this chapter, has multiple meanings.

The first meaning of the phrase 'blue didn't exist' has to do with the wavelength of visible light corresponding to the colour blue. Saying that blue didn't exist in that context would make no sense. Early humans could almost certainly see the colour blue, but they did not name it as a distinct colour.

Which brings me to the second interpretation. The word 'blue' didn't exist. This is closer to what I'm referring to.

Ancient Greeks didn't have a word for blue.

One of the most quoted examples comes from a study of Homer's works, in which there is no mention of the word blue. Some people use this to hypothesize that ancient Greeks didn't see blue, but to me that argument sounds quite similar to claiming that the world existed in black and white before the colour camera was invented.

In fact, most ancient languages did not have a word for blue, with blue being the last colour to be named in most existing languages. When they did, civilizations often – but not always – used the same words to describe both blue and green.

To understand why, we must get into why we as humans name things. Usually, names are words we give to things we have to refer to repeatedly. It's easier to call me 'Aparna' than 'that girl who has a dog and spouts facts randomly'.

So if you were an ancient human, what would you refer to first? Something to differentiate night and day. Check – most societies soon develop colour names that refer to some form of dark and light. Next, something to point at trees and leaves – green is often one of the first colours named. Probably something that points to fruits and berries? Indeed, reds and yellows are next in line. When do the blues get a name? When a society reaches the level of sophistication where jazz bands start playing the blues, I presume.

Oreo's giving me another one of his long-suffering looks, which tells me exactly how much he approves of my humour, so I shall get back to the point. Blue is usually the last colour to be named because objects that are blue are very rare in nature.

Which brings me to the third interpretation of the

statement 'blue didn't exist'. The pigment blue is very rare in the natural world.

In fact, blue is one of the colours for which there are almost no natural pigments. That is why there is a striking lack of blue minerals in nature, and gemstones like blue sapphires and lapis lazuli are extremely precious and sought after.

What about plants then? Surely people had seen blue flowers or blueberries. Again, not quite. Blue is one of the segments of the wavelengths of light that is used by plant cells for carrying out photosynthesis. Blue light is high in energy and is readily absorbed. It's like blue is a special currency that allows plants to make their food.

Due to this fact, blue is seldom reflected by any part of a plant – it's simply too expensive. Blueberries aren't blue – they're purple. They get their colour from anthocyanins, which are the pigments that give fruits, and sometimes leaves, red and purple hues. Carotenoids are the other common plant pigments, giving us yellows and oranges.

Right beside blue in the visible spectrum, though, is the colour green. Chlorophyll – the plant pigment that makes plant leaves and stems green – absorbs blue and red wavelength for use in the energy cycle for the plant. It does not use green much and reflects it away, making our surroundings the medley of green we are used to seeing. So, while we associate the colour green with plants and leaves, it is in fact the colour leaves and plants reflect while absorbing the others. It's one of nature's neat little paradoxes that green, the colour we associate with plants, is actually the colour they don't use.

Speaking of nature's colour tricks, let us talk about animals. Aren't there animals that are blue? Well, there aren't many.

This, again, goes back to where animals get their pigments from. While they don't photosynthesize, many animals get their pigments from what they eat – which for herbivores is plants, and for carnivores is other animals that eat plants.

Let's take the case of flamingos, those bright pink creatures that stand on one leg. Would you believe that in reality these fabulous pink divas of the bird world are actually born grey? Over the course of their lives, they consume a lot of tiny pink shrimp and algae, which gives them their characteristic pink colour. The shrimp in turn also get their colours from eating the red-pink algae, making all of this a pink food chain[5] – algae makes shrimp pink, shrimp makes flamingos pink.

So that's why we don't have many blue animals. Blue pigments are costly to produce, and not easy to acquire through diet. That's the idea behind the phrase 'blue didn't exist' – blue pigments are rare in nature, so earlier civilizations did not have many examples of it.

Okay, but wait. Maybe there weren't any plants or animals that were blue, but the sky always existed, and the ocean was always there. Right?

Well, if you think about it, the sky is not always blue; sometimes, it is overcast and grey. At sunrise and sunset, it's red and orange, and at night, it's black. So again, if you didn't have a name for the colour blue, the sky would just be 'sky-coloured'. As for oceans and rivers, they are reflections of the sky, with rocks and other muddy bits thrown in. The sea would also change colour depending on the time of day,

---

[5] Unlike the so-called 'pink tax', this affects flamingos and shrimp of both genders and all sexual orientations equally.

the time of year and its geographical location. Homer calls it the 'wine-dark sea', for example.

But is it fair to say that blue didn't exist just because humans didn't have a name for it?[6] I'd argue it is. Names are powerful, and when things are named,[7] we tend to identify them better, and faster.[8] In a study where participants were shown some closely related shades of blue, Russians, who have different words for these different shades of blue, processed the differences between the colours much faster than English participants, for whom the colours were represented by a single name – blue.[9] In a similar linguistic quirk, in Japan, the traffic lights are blue-green, and they are called blue (*ao*) by the Japanese, while the word for green, *midori*, is used to refer to things like leaves.

Now, this would all make sense – except, as I told you earlier, the statement 'blue didn't exist!' was completely true but also blatantly false.

Here's a hint for the second part of that statement. I am

---

6 This is a variation on the Zen theme 'Did a tree fall in the forest if no one saw it or heard it fall?'

7 The naming of cats, for example, needs to follow a strict set of rules, as laid out in the poem 'The Naming of Cats' by T.S. Eliot.

8 Again, by we, I mean we as humans. As a species, we tend to be very us-centric when defining most things. Schrödinger's thought experiment asking whether a cat is dead or alive is only valid from our perspective. From the cat's perspective, the answer is much more straightforward.

9 The study can be found here: Jonathan Winawer, et al., 'Russian Blues Reveal Effects of Language on Color Discrimination', Proceedings of the National Academy of Sciences of the United States of America, Vol. 104, No. 19 (2007): 7780–85. DOI: 10.1073/pnas.0701644104.

sitting in India as I write this. How can I say that blue is rare in nature when our national bird, the peacock, struts around wearing the most dazzling blue nature has to offer? Did peacocks have any special supply of special blue food that they hid from other animals?

Well, no. Peacocks, indeed, are one of the rebels of the animal world. They found a loophole in nature's 'no blue pigment' rule through something known as structural colour. Instead of using pigments, they created tiny structures in their feathers that manipulate light itself. It's like they built millions of microscopic prisms that capture and reflect only blue light. Peacocks didn't get the memo about blue being rare – or maybe they did and decided to show off anyway.

Think of it like this – while other animals painted themselves with natural pigments, peacocks built tiny light-bending sculptures. It's the difference between painting your house blue and covering it with specially arranged mirrors that only reflect the blue sky. The peacock tarantula, native to the Andhra region in India, pulls the same trick – no pigment, just perfectly arranged microscopic scales that say 'blue or nothing'.

As we turn into one of the quieter lanes, I realize I've been telling you only half the story. While blue might be rare in nature, India has always had a special relationship with this elusive colour. Long before Europe had a word for blue, India had 'neel'. This is because ancient India used to grow one of the earliest – if not the very earliest – plant-based blue dyes.

The indigo plant contains a molecule known as indican, which oxidizes to the colour blue. So when the leaves of the plant are crushed, fermented and allowed to oxidize, they turn blue. Ancient merchants from India grew and traded this colour across the world, providing an alternative to the extremely expensive ultramarine blue, which is made by crushing gemstones. Here's what I find hilarious – the trade led to another name being generated in English: 'indigo', from 'Indica'. Remember when our dear friend Newton added indigo to the colour spectrum because he wanted it to be divided into seven colours? Inadvertently, long before there were Americans with the abomination that is 'chai tea latte', it was Newton who split the colour spectrum into VIBGYOR putting blue, and indigo (Indian blue) right next to each other.

India, in fact, was one of the oldest civilizations that had a name for the colour blue. The colour was important in religious and social contexts. Lord Krishna, for example, is often represented with blue skin. Some say it represents his infinite nature, like the boundless sky, while others point to ancient texts describing his skin as the colour of rain-laden clouds.

This association of blue with royalty and luxury has deep historical roots. In ancient Egypt, which was another civilization with a name for blue, it was the colour of the gods and the pharaohs. Cleopatra, the last active ruler of the Ptolemaic Kingdom of Egypt, was known for her love of the colour blue. She used powdered lapis lazuli, a deep blue semi-precious stone, as eyeshadow, a cosmetic choice that was both a display of wealth and a nod to the divine. Talk about a bold fashion statement! The Egyptians also used blue in their jewellery and decorative arts, often in the form of faience, a glazed ceramic that could be tinted various shades of blue – because nothing says 'I'm a god-king' like accessorizing with blue ceramics.

Speaking of religious blue, here's a plot twist that would make any storyteller proud. In early mediaeval Europe, blue was actually considered the devil's colour. Early Christian churches saw blue's rarity as something suspicious – after all, what godly hue would be so difficult to obtain? Demons were often painted in blue hues, marking their otherworldliness. An example is Satan being depicted as a blue angel in a sixth-century mosaic from the Basilica of Sant'Apollinare Nuovo in Ravenna, Italy.

But then, around the eleventh century, something remarkable happened. The church pulled perhaps the greatest rebranding move in colour history – they began depicting the Virgin Mary in brilliant blue robes, usually painted with ultramarine pigment made from crushed lapis lazuli. The same rarity that had made blue suspicious now made it precious enough to be worthy of the divine.

This transformation from demonic to divine wasn't just a religious shift – it was an economic and artistic revolution. Patrons started to specify in artists' contracts exactly what kind of ultramarine should be used in Mary's robes. The more expensive the blue, the more devoted the patron was considered to be.

Indigo was the alternative blue pigment, and had the nickname 'blue gold'. While the pigment was a rare commodity all over Europe, we had the whole city of Jodhpur or 'Neel Nagri', with large sections of the old city painted in striking shades of indigo. One of the theories is that the indigo mix in the whitewash helped keep houses cool and repelled termites, while other theories suggest that it started with the

Brahmins painting their houses blue to differentiate them from the rest. Yet another theory suggests that it was just the aesthetics of the colour that led to the entire city adopting it. The reality, perhaps, is a mix of several sociological, scientific, aesthetic and practical variables.

What the aforementioned facts tell us is that whether in life or in linguistics, most things are rarely just black and white. Even black and white themselves (or any other variation of dark and light) are followed by reds, greens, yellows and finally blue, in our linguistic patterns.

India's relationship with blue also has a darker side. During the British Raj, the East India Company, seeing the enormous profits in trading the blue dye, forced farmers to grow indigo instead of food crops. The farmers would take out loans to plant indigo, but the British would set prices so low that they could never repay their debts. Entire families were trapped in cycles of debt, violence and death. The fertile plains of Bengal, which should have been glowing with golden rice, were instead blooming blue with oppression.

The exploitative nature of indigo farming gave India its own case of the blues. Indigo farming was harsh both for the local environment and the farmers. While repeated indigo farming depleted the essential nutrients from the soil, the extraction process itself was brutal – workers stood for hours in vats of fermenting indigo, developing painful skin conditions and respiratory illnesses from the chemicals involved.

Colonial impacts on agricultural practices aren't just a thing of the past, and we need to remember the past to make sure that we don't repeat it. One of the early patriotic works of Indian literature is *Nil Darpan* (1860). This is a Bengali play by Dinabandhu Mitra, centred on the Indigo Revolt of 1859 when farmers in Bengal rose in protest of the enforced cultivation of indigo.

The play reflected the realities faced by the indigo farmers – intimidation, exploitation, violence (including sexual violence) at the hands of planters and lack of redress through the judicial system. When translated into English, it raised awareness and shifted public opinion in England against the indigo system. Here's the irony: James Long, who published the English translation, was taken to court by the proprietor of *The Englishman* newspaper, together with the Landholders Association of British India and a general body of indigo planters, for libel. He was found guilty, sentenced to one month in jail and fined 1,000 rupees.

There was some respite. The revolt and the eventual public opinion shift led to the British government establishing the Indigo Commission in 1860 to investigate the abuses, which in turn forced temporary concessions through the Indigo Contracts Act – a small victory carved out of a much larger injustice.

Yet the system held firm. The violence continued. Until eventually, it didn't.

Indigo didn't fall out of favour because of moral reckoning. It ceased to be a valuable commodity because Europe no longer needed it.

Industrial production of blue pigment had started as early as 1704, when a German paint maker named Johann Jacob Diesbach accidentally created a pigment now known as Prussian blue. Its deep, rich hue and easy availability made it a favourite among artists, who could now access a wider range of blue shades without relying on expensive natural pigments like ultramarine or indigo.

However, Prussian blue was relatively unstable, and indigo continued to be used to dye textiles and for other industrial uses.

Then in 1897 came the final blow, in the form of Adolf von Baeyer and the German company Badische Anilin- und Sodafabrik's (BASF) synthetic indigo. A blue that didn't need seasons or sunlight – just factories.

Suddenly, indigo lost the quality that made it precious – its rarity. Slowly, the crop that had once ruled Bengal through violence and control was cast aside. The same imperial powers that had demanded its growth, punished farmers unwilling to grow it and profited from its harvest now had no use for it. Indigo had become inconvenient. And so the blue terror came to an end.

But the damage was done. The soil remembered. So did the bodies. So did the silence.

Blue has always carried contradictions. It has been both sacred and brutal, desired and discarded. It has appeared in military uniforms and mourning clothes, on currencies and canvases. It has represented royalty, rebellion and ruin.

In many ways, the story of India and blue is similar. It is simultaneously beautiful and painful. It is both untold and unfinished. And, in a cosmic twist of fate, blue is what the Indian cricket team adorns while dominating the global stage in the game of cricket – a game brought to us by the British.

We've since then transitioned into a world where the colour blue is everywhere. Artificial pigments that correspond to the hue of blue are plentiful, and their discovery has revolutionized the way we perceive and use this once-rare colour. Prussian blue, which is believed to be one of the first artificial blue pigments, set the ball rolling, dethroning the rare ultramarine blue that artists strived for.

The kicker? Johann Jacob Diesbach aka our aforementioned paint maker in the early eighteenth century, wasn't trying to make a blue pigment. While attempting to make a red pigment, he used potash (potassium carbonate) that had been contaminated with animal oil. Instead of red, this combination unexpectedly gave him the colour everyone wanted – bright blue. Prussian blue quickly became a game changer in the art world.

The impact of Prussian blue extended far beyond the realm of art. Its widespread use gave rise to the term 'blueprint', which originally referred to the process of reproducing technical drawings using a Prussian blue photographic print. The term has since evolved to mean a detailed plan or design, a testament to the pigment's influence on our language and thinking. Who knew a colour could be so bossy, telling us how to plan and execute our projects?

Artists and their fascination with blue continues throughout time. There is a painting by French artist Yves Klein: It is a square of blue colour and is often used as an example of the 'my five-year-old could paint that' style of paintings. However, the story behind the painting is what makes this piece of art powerful. Remember how Vantablack was a completely new type of black that didn't exist before? Well, Klein did something similar with blue – back in 1960. He created his own unique shade of blue, called International

Klein Blue (IKB), which didn't exist before. That is why the painting is considered remarkable. It was the first time that shade of blue appeared anywhere. If a five-year-old did manage to do the same, I'd be pretty darn impressed!

Today, blue is ubiquitous in our lives. It's in the clothes we wear, like the quintessential blue jeans that have become a global fashion staple. Interestingly, the popularity of blue jeans can be traced back to India, from where the sturdy blue fabric called 'dungri' was exported to Europe and America. The indigo-dyed cloth caught the eye of Levi Strauss, who used it to create the first pair of riveted denim jeans. So, the next time you slip into your favourite blues, perhaps you'll have a chuckle to yourself remembering that one of the biggest reasons these jeans that you hold today exist is because some guy named Levi thought it would be a good idea to rivet pants made from Indian fabric.

Fashion is a strange beast. Meryl Streep's iconic dialogue from *The Devil Wears Prada* – 'It's sort of comical how you think that you've made a choice that exempts you from the fashion industry when, in fact, you're wearing a sweater that was selected for you by the people in this room … from a pile of "stuff"' – also features the diva that is the colour blue. In fact, it is not just blue, it's not turquoise and it's not lapis – it's actually … cerulean.

Blue also dominates our digital world, from the 'calming'[10]

[10] Blue light at night is thought to disrupt our natural rhythms, impeding sleep and recovery. Yet in the modern world, we are surrounded by it in our big screen we often use till late at night, and the smaller hand-held screen we all love to scroll right before bedtime.

hue of our computer screens to the logos of tech giants like Facebook and Twitter. The granddaddy of tech – IBM – is often called Big Blue because of the colour of its logo.

In the realm of advertising, blue is often used to evoke feelings of trust, reliability and tranquillity. Many brands, particularly in the finance and healthcare sectors, use blue in their logos and marketing materials to create a sense of stability and professionalism.

Luxury brands, too, have embraced blue, perhaps as a nod to the colour's historical association with rarity and prestige. After all, for centuries, blue pigments were some of the most expensive and sought after, reserved for the robes of royalty and the most opulent of artworks. By using blue in their branding, these companies are tapping into that deep-seated sense of value and exclusivity. It's like they're saying, 'Hey, remember when only the super-rich could afford blue? Well, you can also be a part of this super-exclusive-elite-rich-and-successful group if only you bought this thing we're trying to sell. Isn't that neat?'

Oreo stops to investigate what must be a fascinating message on a car tyre. For dogs, car tyres are the equivalent of social media. Cars travel all over the place and dogs do things on tyres everywhere. Since smell based cues last a relatively long time, a dog may leave a 'message' in the morning, which can be read by another dog in the afternoon, who will then leave his own reply to the 'message'. This sets up a cycle where other dogs find the messages fascinating and do more things on tyres, creating long message threads aka dog social media. So as Oreo checks in on his social media for the day, I wonder what colours he sees in our morning world.

People often say dogs are colour-blind, but that's not quite right. While I'm admiring the fading purple of a morning glory creeper draping over a compound wall, Oreo sees the world in a different palette altogether.

'What do you see, boy?' I ask him as he thoroughly investigates a patch of grass. Dogs have only two types of cone cells in their retinas, compared to our three. While we see the world in red, green and blue wavelengths, Oreo's world is primarily blue and yellow. It's similar to what a human with red-green colour-blindness experiences.

This makes me smile – in a way, Oreo is better equipped to appreciate the rare blue in nature than I am. While I'm distracted by every shade in the spectrum, he sees the world in more subtle contrasts. The bright red collar of that indie dog across the street? To him, it's probably a muted yellowish grey. But the brilliant Bangalore blue sky? That he sees almost as clearly as I do.

We pass by one of the old rain trees that line our street, its branches reaching out like gnarled fingers above us. These colonial-era giants still provide patches of shade that make our morning walks bearable. The relatively sparse traffic in this area means you can still hear birds – not the symphony you'd get in a dense green forest, but the persistent cawing of crows, the occasional barbet and, if you're lucky, a golden oriole flashing through the gaps between buildings.

Oreo suddenly alerts me to something I can't perceive – probably one of the neighbourhood cats. His colour vision might be limited, but his other senses more than make up for it. Dogs (and cats) can see better in dim light than humans, thanks to a special reflective layer behind their retinas called the tapetum lucidum. That's what makes their eyes glow in photographs. They also have a wider field of view than humans, about 240 degrees compared to our 180 degrees. What they lose in colour discrimination, they make up for in motion detection and night vision.

Speaking of night vision, here's a fun detour. That trope about carrots being good for your eyesight? It was propaganda spread by the Allies during World War II. During the aerial Battle of Britain in 1940, the British relied on a top-secret invention – radar on board their fighter planes to guide their pilots in the dark. In order to conceal the existence of this onboard radar, they popularized the idea that carrots were incredibly good for night vision. Like much propaganda, it has a kernel of truth – carrots are good for our vision, but so are other fruits and veggies, and carrots don't bring anything that special to the food table.

'Watch the bump! Jump!' I tell Oreo as we navigate around a tree root that has created a mini-mountain in the footpath. Oreo has his favourite spots mapped out – that one corner where frangipani flowers collect; the compound wall where someone's bougainvillaea spills over in what I see as a riot of purple, but what he probably perceives as various shades of blue-yellow. A butterfly lands on a flower, capturing Oreo's attention.

I wonder if an increased susceptibility to motion is why he's so fascinated by the local butterflies – it's not their colours that catch his attention, but their movement. Oreo also firmly believes that if only I would unleash him, he would jump up and catch one of those kites flying high up in the sky. It's probably my fault. I didn't quite explain that this isn't what I meant when I told him to reach for the skies.

As it stands, when I look up, the blue of the sky is broken by the browns and greens of ancient trees. Between these ancient trees, younger gulmohar and copper pod trees add their own splashes of colour – rebellious oranges and yellows against the urban landscape. This of course reminds my brain that if you eat a ton (like really an abnormally high amount) of orange foods like carrots and pumpkins, your skin – if it tends towards a lighter undertone – may turn a subtle shade of orange. Not orange like a traffic cone, but more of a subtle afterglow. For people whose skin is darker, such an effect would not be visible because of colour theory. I chuckle at the thought of turning orange but remember how much I'd have to eat to make a difference. So please, eat your veggies anyway!

To Oreo, these vibrant flowers might all look similar shades of yellow, but he doesn't seem to mind. He's more interested in the stories told by scents than by colours anyway. While I'm marvelling at nature's palette, he's reading a complex narrative of who's passed by, what they had for dinner and probably their entire life story, all from a single sniff.

Dogs smell in rich textures and 'colour' compared to us humans, since they have millions more in the way of receptors in their noses. They can differentiate smells we humans may not even perceive. There's disorder and duality in many things in nature, and one way to understand the differences is by taking help from our fellow travellers on this planet to give us a glimpse from a different perspective.

As we head further, I think about how relative our perception of colour is. Here I am, fascinated by blue's rarity in nature, while my constant companion experiences a world where blue is one of his primary colours. Maybe that's why he's so unimpressed by my colour theory musings – to him, blue isn't rare at all. It's just another part of his dichromatic world, no more special than the yellows that make up the other half of his colour spectrum.

'You know what, Oreo?' I say. 'Maybe that's why we're so fascinated by blue. It's the colour of mystery, of depth, of things just beyond our reach.' He responds by giving me a look which probably means he thinks nothing of my philosophical conclusions and would like me to move a bit faster. I ignore him and continue on my musings.

From a colour that was once so rare that it didn't even have a name, to a pigment that was prized above all others, to a hue that's now so common we barely give it a second thought. The next time you see blue, whether it's in a painting, on a screen or in a pair of jeans, take a moment to appreciate its journey. From the ancient Egyptians who ground lapis lazuli into powder for Cleopatra's eye shadow, to the Indian indigo farmers who toiled under the oppressive thumb of colonialism and the modern-day scientists who have created new shades of blue that our ancestors could never have imagined – each of them has played a part in the story of this remarkable colour.

Who knows what the future holds? With new pigments and technologies emerging all the time, the possibilities for blue are endless. One thing is for sure, though – our love affair with this most enigmatic of colours is far from over. Blue, it seems, has captured our collective imagination like no other colour, and its hold on us shows no signs of loosening. From the clothes we wear to the screens we stare at, from the logos we trust to the sashes of royalty or the jerseys of Indian cricket players, blue is everywhere, and everywhere, it is loved. And why not? After all, it's the colour of kings, queens and overpriced eye shadow. What's not to love?

A squirrel runs in front of us out of the blue, and I get pulled out of my philosophical musings by a dog who suddenly catapults forward, galvanized by a much more urgent matter at hand.

3

It's a Bird, It's a Plane, It's a Small World (Network)

Follow
Someone
Home

I chuckle at the traffic sign. Someone at the Bangalore Traffic Department has a somewhat creepy sense of humour.

Anyway, Oreo and I continue on our 'walk', which is more of 'pulling on the leash to run after squirrels', and suddenly 'stopping to sniff a rock … just because', rather than a dignified, steady progression. The squirrel we were chasing in the last chapter pulled off an escape manoeuvre by climbing on a tree. So here we are, back to our regular programme, though a sulking Oreo is convinced that if only I had allowed him to, he would have caught that squirrel, trees or no trees.

Closer to the signage, I note that it actually reads:

Follow traffic rules, Someone is waiting at Home for you.

Only, much of the text is written in such a small font that it's only visible if someone squints hard at it.

My mind segues from that sign to parallels in the world around us, with its plethora of things that are hidden at first glance. It takes someone with an unusual perspective – a scientist, an artist, a philosopher or perhaps a dog – to stop and look carefully enough to see the unseen elements in our ecosystem.

My walks with Oreo often help me decipher the seen and the unseen. We often take routes that would not make sense at first glance. We avoid straight paths, instead taking winding routes and smaller bylanes which make the distance between points A and B longer than needed. But then, our walks are not about getting from one point to another as quickly as possible but about systematically patrolling and inspecting the neighbourhood that Oreo considers his territory.

Oreo looks at me, trying to communicate the salacious gossip he has just learnt by sniffing the lamppost. Unfortunately for him, I don't understand the thoughts he is trying to project into my head like a Jedi master. Even if he could put the information in words, the way his nose interprets the world around us is probably too different from mine for me to understand. It's like he's communicating via interpretive dance, while I have two left feet. Dance, incidentally, is another form of movement where the goal is not to get from A to B, and if we judged salsa on the efficiency of transport, it would also come up severely short.

Often, the right – or rather, different – perspective helps us see a whole new layer of what is going on beneath the surface, just as using the right names can help us see various colours and shades of colour. As Oreo pulls on the leash, eliciting a stern look from me, I look at the back of my hand, and my anatomy knowledge kicks in. I am transported back to my biology lessons, where my very enthusiastic teacher taught us about the networks within our bodies.

He took us on a journey discussing how multifaceted the circulatory system is. He told us about the arteries and veins that transport blood across our body, and how this is a simple yet incredibly complex network. We have a heart which, as Leonardo da Vinci said, is formed like a vortex, pumping blood 24x7. The blood carries essential nutrients and oxygen to our organs, pulsing through the arteries, while at the same time carrying waste products away from organs. It is blood that carries specialized cells fighting bacteria and other pathogens, and it is also blood that rushes to your cheeks to show everyone when you're embarrassed. The things being transported here are incredibly small. Yet, inside each of the trillions of cells that make up our bodies, a complex transportation network is constantly at work. Nutrients, proteins, waste products – all need to be moved to where they're needed or disposed of, and our circulatory system is the network they all utilize.

Our teacher told us to remember this as we headed to the lab for our practicals. There we looked under the microscope to see the blood cells, and marvelled at how what we assumed to be a simple liquid actually contained so many different types of cells. Of course, as almost always happens in biology labs, a student fainted at the sight of blood.

I'm pulled back to reality by a tug at the leash, and my stern look is returned by Oreo, who is done sniffing this spot and would like to move on to the next one as soon as possible, please and thank you!

'Right then, let's plot our course,' I tell Oreo, who responds with a look that clearly says, 'You humans make everything needlessly complicated.' He's not wrong. While I'm mentally calculating trajectories and trying to figure out which roads have mysteriously changed direction since yesterday, Oreo is processing spatial information in a way that would put most supercomputers to shame.

Let's start with one of Oreo's superpowers – his nose. In a dog's brain, scent signals don't travel in straight lines from the nose to the processing centre. Instead, they follow what mathematicians call a 'small-world network' pattern. It's like the canine nose plays an elaborate game of pass the parcel with scent molecules.[1]

When Oreo sniffs a lamppost, odour molecules bind to receptors in his nasal cavity, triggering a neural conga line. These signals zip through his olfactory bulb[2] and into the brain's olfactory cortex faster than you can say 'squirrel'. His brain's small-world network allows for rapid, parallel processing of different scent components.

'Oreo, let's back up for a bit,' I tell him as I realize that our typical route has been dug up yet again, and we need to go back and take a different route. Which reminds me, I should go back and talk about small-world networks before telling you how a neural network is an example of one.

Now, what is a 'small-world network'?

It's a fascinating story that starts over 50 years ago.

Imagine, if you will, a world before the Internet, before social media, when the idea of being connected to anyone on the planet seemed like science fiction. This was the world in the 1960s, when a social psychologist named Stanley Milgram

---

[1] Dogs have been used for sniffing out everything from drugs to cancers, thanks to this superpower.

[2] Olfactory centre and cortex are the fancy scientific terms to describe the smell-processing centre; 'Smell-processing centre' is apparently too vague and broad for scientific discourse. Blame the Greeks!

decided to play a giant game of 'pass it on' with the entire United States.

Milgram was fascinated by a simple question: In this vast world, how connected are we really? It's the kind of question that might pop into your head while standing in a crowded market, wondering how many degrees of separation exist between you and, say, the president of India.

So, Milgram devised an experiment that was as simple as it was ground-breaking. He sent out packages to random people in the Midwest (which consists of 10-odd states in the middle of the Continental United States of America), with instructions to get these packages to a specific person in Massachusetts. The catch was that they could only send their packages to someone they actually knew on a first-name basis.

If you want an Indian equivalent, this would be like picking six people from Delhi at random, and asking them to deliver sweets to a resident of Tuticorin who was also picked at random. The Dilliwallahs could only send the sweets to people they knew personally, who could in turn only send the sweets to people they knew personally. Assuming nobody ate them, how fast would the sweets be delivered?

Now, you might think these packages would end up bouncing around the country like a game of postal pinball. But here's where it gets interesting: Most of the packages that made it (and yes, some got lost) did so in just about six steps.

This experiment is also what helped popularize the famous idea of 'six degrees of separation' – the notion that you could get from any person to any other person in the world through about six connections of only people who know each other.

Fast forward to the late 1990s. Two mathematicians,

Duncan Watts and Steven Strogatz, were pondering this phenomenon. They were like two kids staring at a giant ball of tangled yarn, trying to figure out how it all connected.

You see, before Watts and Strogatz came along, mathematicians and scientists were trying to understand complex networks using two main models.

First, we had the 'regular' networks. These were neat, orderly and about as exciting as a perfectly arranged sock drawer. In mathematical terms, these networks had a high clustering coefficient (meaning your neighbours were likely to be neighbours of each other) but a large average path length (it took many steps to get from one side of the network to the other). Imagine a city where you know all your neighbours, but to get to the other side of town, you have to visit every single house along the way. Not very efficient, right?

On the other flip-flop, we had 'random' networks. These were the wild children of network theory, with the connections here made willy-nilly. Mathematically, they had a low clustering coefficient (your neighbours were unlikely to know each other) and a small average path length (you could get across the network quickly). Think of a city where you can teleport to any random house, but you don't know any of your neighbours. Fast, sure, but not very realistic.

The problem was, real-world networks didn't behave like either of these models. They were like that perfect cup of chai – neither too rigid, nor too random, but just right.

Enter Watts and Strogatz, stage left. They proposed a model that started with a regular network, and then, like a mischievous child rearranging the sock drawer, they added a few random connections. The magic ingredient? A parameter they called 'p', which determined the probability of rewiring each edge of the network.

Here's what they found out. When p was zero, the network could not be rewired. It remained like the extremely well-ordered sock drawer, as the randomness was zero. When you maximized the value of p, the network became completely random, with low clustering and short paths.

But when p is *juuuust* right (a small value that is much less than 1, but more than zero), something magical happens. In this Goldilocks zone, the network maintains a high clustering coefficient (so your neighbours still know each other), but the average path length drops dramatically. Suddenly, you can get from one side of the network to the other in just a few hops.

Okay. We are going to delve into a little bit of math. Stay with me here.

Mathematically, Watts and Strogatz's model involved two equations that can be interpreted as follows: There's a clustering coefficient 'C' that describes the closeness between neighbours – the larger the value of C, the closer things are clustered, and that's good for network connectivity.

That clustering coefficient changes in value as the network adds random connections. The change in the clustering coefficient C(0) from the initial value (which is high since the first two persons in a Milgram-type experiment know each other) to a new value C(p) depends on a relationship defined by the 'rewiring probability' p as new connections are added.

The initial value C(0) is multiplied by the factor $(1 - 3p^2/4)$ to derive the new value C(p) as connections are added. Basic algebra indicates that if the value of p^2 is very small – much less than 1 – the clustering coefficient C(p) will remain high as connections are added.

This first equation defines the value of the clustering coefficient:

$C(p) \approx C(0) \times (1 - 3p^2/4)$, where ≈ means 'approximately equal'

The second key equation defines average path length:

$L(p) \propto \log(n)/\log(np)$

This describes how the average path length L(p) between nodes shortens logarithmically as rewiring probability increases. Here 'n' represents the total number of nodes in the network (the number of people who pass the parcel in Milgram's experiment), and the symbol '∝' indicates direct proportionality. The logarithm of n (the natural logarithm to the base e if you're a nerd like me) is divided by the logarithm of n into p (the rewiring probability).

We want the path length L(p) to be as short as possible, which means that the quotient or result of this division should not be large. Hence, log (np), the divisor, should be relatively large compared to log (n). This happens when the rewiring probability p increases.

So, one of these equations should ideally have a high rewiring probability to shorten the average path length L(p) while the other should have a low rewiring probability to ensure that the clustering coefficient C(p) is high!

The Goldilocks value is seen when the p is large enough to reduce path length without being so large that clustering drops too much. That's when the 'small-world' effect happens (when $p>0$ and $p<<1$, if you're into the details) and nodes become connected through fewer steps despite maintaining relatively high clustering.

Together, these two equations describe how clustering and path length change as random rewiring (parameter p) increases in Watts–Strogatz small-world networks.

So what does all these symbols and math mean? With this model, we could see how large world networks, can become connected without adding many steps.

The real kicker? The transition from 'large world' to 'small world' happens very quickly as p increases, and it happens while the network still looks mostly 'regular' to the naked eye. It's like adding just a few strategic shortcuts to our imaginary city, and suddenly everyone's connected with everyone else. Think of a city's public transport system: It's not just individual bus routes linking neighbourhoods but also a whole network of interconnected routes and transfer points, acting like shortcuts to help commuters get from one side of the city to the other.

These, my friends, are 'small-world networks'.

The beauty of the Watts–Strogatz model is its simplicity and its effectiveness. With just one parameter p, they described a whole range of networks, from regular to random, and showed how real-world networks often sit in that fascinating middle ground.

Since then, network scientists have built on this foundation, developing more sophisticated models to describe specific types of networks. But the small-world model remains a cornerstone of network theory – once you get it, every sort of network clicks.

Remember earlier how I told you that a different perspective can show you things hidden in plain sight? Well, once the model was developed, we started finding it everywhere!

Small-world networks broadly[3] explain how your brain, or Oreo's brain, with its billions of neurons, can process information so quickly. Your neurons aren't all connected to each other (that would be a real mess), but they're not randomly connected either. They're in this sweet spot of small-world connectivity.

The power grid? That's a small-world network too. It's why a squirrel chewing on a wire in Bangalore can theoretically cause a blackout in Belagavi.[4]

Even the spread of infectious diseases follows small-world network patterns. It's like how one person sneezing in a Bangalore bus can somehow lead to half the IT Park calling in sick the next day, or how one infected person at a conference can end up giving Covid to over a quarter of a million people. But let's not dwell on that – we've all had enough pandemic talk to last us through several small worlds.

Now, here's where it gets as exciting as finding an empty road during peak hours[5]: The math behind small-world

---

[3] No mathematical theory exactly replicates real world neural connectivity, but small-world networks, or models based on small-world networks, can act as a good approximation for broad trends.

[4] Note: No squirrels were harmed in the making of this analogy, though several were spotted looking hungrily at power lines.

[5] I know that's impossible, but go along with it!

networks is incredibly useful for solving complex problems. It's like having a Swiss Army knife for data scientists and network theorists. Imagine if you had a tool that could optimize Bangalore's traffic flow, predict the next viral meme and also tell you which samosa stall has the shortest queue – all at once. That's the power of small-world network analysis.

Do you want to optimize supply chains? Small-world network analysis can help find the most efficient routes, ensuring your online shopping arrives before you forget you ordered it. It's like having a delivery person with the navigation skills of a migratory bird and the speed of a Bangalore techie rushing to make it to a 9.00 a.m. meeting.

Are you trying to understand how information (or misinformation) spreads on social media?

Small-world networks explain why that one post from your aunt about a miraculous weight-loss tea can suddenly go viral, reaching more people than your carefully crafted research paper ever did.

'Oh hey! Sorry!' I apologize to a delivery rider as Oreo decides he needs to sniff the tyres of his bike – while it's moving. The rider ignores me and keeps driving towards his intended location, perhaps to deliver the potatoes in his bag to the customer in under 10 minutes.

Ten-minute deliveries – they sound and feel like magic, no?

As I mentioned a little bit earlier, the magic ingredient there? Small-world networks.

Unknown to you, your simple act of tapping 'Order Now' sets in motion a complex dance of algorithms and real-world logistics that would make even the most intricate Bollywood choreography look simple. Behind the scenes, in a nondescript building just a few blocks away, your order pings on a screen. This isn't your regular kirana store; it's a 'dark store' – a secret agent of the delivery world, invisible to the public but crucial to the operation. And just like secret agents, their utility lies in being undetected but omnipresent.

Anybody who's looked at the organization inside a traditional kirana store knows that they are anything but 'traditionally categorized'. Kirana stores are driven by the shopkeeper's unique data insights about customer preferences and optimized 'laziness'. The shopkeeper and his assistant Chhotu don't want to move around too much to get you the specific item you want. Commonly sold stuff will be sitting on a shelf behind the shopkeeper's chair, but that specific perfume that only a few people want is probably in the storeroom at the back or shelved up high with the shopkeeper only getting it when asked, probably accompanied by a grumble.

But while the kirana shopkeeper only has the data pertaining to local residents' tastes and preferences, dark stores can crunch much more data. Inside this dark store is a space where every item has its perfect place,[6] determined not by traditional categorization but by an intricate web of data-driven decisions. That packet of masala peanuts you ordered? It's not sitting next to other snacks. No, it's strategically placed next to the cold drinks because the all-knowing algorithm has determined that these items are often ordered together.

As your order comes in, a picker springs into action. They

---

[6] Maybe not perfect overall – more like 'perfect', as detected by the data of today.

move through the store with the precision of a dancer and the speed of a sprinter, or so the management gurus tell us (the pay is low and the work is tiring, so reality doesn't always match the textbook). The route they take isn't random – it's a carefully optimized path, designed using principles borrowed from small-world network theory. Each turn, each grab, each step has been calculated to ensure maximum efficiency. It's as if the store itself is a physical manifestation of a complex network, with popular items acting as hubs, connected by the invisible threads of customer preferences.

Meanwhile, out on the streets of Bangalore, a network of delivery partners pulses like a living organism. Each rider is a node in this vast, dynamic network, their positions and routes constantly shifting and being optimized. The system doesn't just see them as individual riders[7] but as part of an interconnected web, much like how your brain's neurons work together to form thoughts.

As your picker assembles your order in record time, the system is already calculating the optimal route for delivery. It's not just looking at the shortest path – oh no, that would be too simple. It's considering traffic patterns, weather conditions and even the performance history of available riders. It's like a grand game of chess, where every piece on the board is constantly moving.[8]

---

[7] How that connects to the humanity of the rider is a whole different conversation.

[8] And it's doing this in Bangalore or other extremely congested cities, where traffic conditions and one-way streets can sometimes stretch a 10-minute walk into a 40-minute drive!

Within minutes, your order is packed and handed off to a delivery partner. Back in your living room, you're just about to start the next episode of your binge watch when your doorbell rings. As you open the door to receive your perfectly chilled drink and crunchy masala peanuts, you're not just getting a delivery – you're experiencing the culmination of advanced network theory, data analysis and a good dose of Indian ingenuity aka *jugaad*.

This entire process, from the moment you tapped your phone to the moment your snacks arrived, is a testament to the power of small-world networks. It's turned the sprawling, chaotic city into a finely tuned delivery machine, where every store, every item and every rider is a node in a vast, interconnected web.

Of course, a key tool that is foundational to these rapid delivery systems is GPS-based accurate maps that help the rider navigate to your house smoothly. Today, it is difficult to imagine navigating anywhere without directions on your phone. Every time a friend asks me to meet for dinner, the next message is always 'send the location'.

It wasn't always this way, though. When Google Maps was first rolled out, it encountered a fascinating puzzle in countries like India – a puzzle that would ultimately make its technology much more robust and versatile.

Time for another story! Picture yourself in a narrow alley in Old Delhi or a back street in Tokyo. You're surrounded by a maze of unnamed lanes, each barely wide enough for a car to pass through. Now, imagine trying to give someone directions to find you: 'Take the third right after the temple with the blue door, then left at the chai stall with the orange cat.' Sounds familiar?

This scenario was Google Maps's nightmare in its early days. In many parts of India and Japan, side streets often don't have official names, and even if they do, nobody knows them. Instead, locals navigate using landmarks, relative directions and a mental map built over years of experience. It's a system that works beautifully for humans but poses a significant challenge for algorithms.

To compound the problem, GPS accuracy on early smartphones wasn't always reliable, especially in dense urban areas with tall buildings. Your blue dot on the map might show you confidently striding through someone's living room when you're actually on the street.

Google's solution? They had to think outside the box – or, in this case, outside the street grid.

First, they developed a system to generate unique codes for locations, even if they didn't have a formal address. This system, which later evolved into what we know as Plus Codes, allows every 3 x 3 m square on Earth to have its own 'address'. It's like giving every spot on the planet its own postal code. While this was designed as a simple solution for one problem, it was later used as the basis for many technological innovations, showing how a solution to one problem may lead to solving a whole bunch of other problems.

Next, they turned to the power of crowdsourcing. Google encouraged users to add information about unnamed streets, local landmarks and points of interest. This not only filled in the gaps in their maps but also captured the local knowledge that's so crucial for navigation in these areas.

They also improved their GPS accuracy by combining

satellite data with other sensors in smartphones, like accelerometers[9] and compasses. This allowed them to better estimate a user's position, even when the GPS signal was sketchy.

Perhaps the most innovative solution, though, was the development of visual positioning technology. Google started collecting street-level imagery (think Street View, but more comprehensive) and using machine learning to recognize visual landmarks.

These challenges and their solutions had a profound impact on Google's mapping technology. The need to navigate in places like India, Brazil and Japan pushed Google to develop more flexible, robust systems that could handle a wide variety of environments and naming conventions. The result? A navigation system that's more adaptable and intelligent than ever before. It can guide you through the winding alleys of Chandni Chowk just as easily as it can direct you down the broad avenues of New York City.

Moreover, these innovations have applications far beyond just helping you find the nearest coffee shop. The ability to precisely locate and navigate to any point on Earth, even without a formal address, led to advances in many fields, covering everything from emergency services and delivery logistics to urban planning.

I pull at Oreo's leash and lead him to a different spot, where he starts to dutifully sniff the dirt, acting as if he didn't just jump into the air like a superhero trying to take flight, except

---

[9] An accelerometer is a sensor that can detect motion, while a compass uses magnetic forces to determine the direction in which you're moving.

his supervillain was the tyres that deigned to move too fast when he had decided to sniff them.

The traffic in front of my house is a mess. While we have the best technology helping us navigate, the traffic is at a standstill. Everyone is in a rush, so nobody moves, showcasing the wonderful behavioural quirk of 'It's okay if no one moves, as long as I am the one to move first when we do'.

This all seems in stark contrast to ancient trade routes. No Internet, no GPS, and yet somehow, things and people made their way to where they needed to go. The secret sauce was – you guessed it – small-world networks.

As Oreo jumps into a neat little pile of dry leaves, the crunching sound as he steps on the leaves making him ecstatic, let me take you back in time.

Imagine a merchant in Chang'an (modern-day Xi'an) playing an elaborate game of pass the parcel with his silk. He'd hand it off to a trader heading west, who'd then pass it to another in Samarkand and so on. This silk would change hands faster than a hot samosa at a cricket match, creating a vast network that spanned continents. It's like a game of Chinese whispers, but instead of mangling messages, they were furthering cultural exchange.

These traders weren't just moving goods; they were also unofficial ambassadors, spreading ideas faster than Bangalore traffic spreads road rage. They carried not just silk and spices but also art, religion and, yes, recipes. It's why you can find variations of biryani from Bangalore to Baghdad, each with its own local twist. It's like the ancient world's version of fusion cuisine, but with more camels and fewer food bloggers.

Now, let's zoom out and look at the big picture – or should

I say, the big tapestry. The Silk Road wasn't just one road, but a complex web of routes that would make Bangalore's maze-like streets look like a straight line. It stretched from China to the Mediterranean, with more branches than a banyan tree in Lalbagh.

But wait, there's more! Long before Marco Polo played the world's most expensive game of 'I spy' along the Silk Road, there was the Royal Road of the Persian Empire. Imagine a highway system so efficient that it puts our modern courier services to shame. A message could travel the entire 2,600-km length of the road in just seven days.

The subject of what was being delivered is another matter altogether. Interestingly, some of it was recorded by the Persian Empire as customs data, which helped in the deciphering of lost Assyrian languages.

Silk road has captured our collective imagination ever since. Incidentally, Marco Polo's book was a bestseller for centuries because the Venetian merchant had recorded conversions of the diverse weights and measures used along the Silk Road.

Yet it wasn't the only ancient trade route of significance in our small-world networks. The Golden Road, connecting the Baltic Sea to the Black Sea, was another key player in this global game of connect-the-dots. Amber, the fossilized tree resin that's basically nature's time capsule, was the hot commodity here. It travelled from the north to the Mediterranean faster than gossip spreads in a Bangalore apartment complex.

Now, you might be wondering, 'What's left of all this ancient hustle and bustle?' Well, quite a lot actually! The Ajanta and Ellora caves in Maharashtra are like a stone Instagram feed of ancient trade influence. You've got Buddhist, Hindu and Jain motifs all hanging out together like different startups in a Koramangala co-working space.

Ever been to Hampi? Those magnificent ruins aren't just a backdrop for your next profile picture. They're a testament to the Vijayanagara Empire's participation in global trade networks. Archaeologists have found Chinese porcelain and coins from Arabia there. It's like finding a Pizza Hut receipt in your Biryani order – unexpected but proof of global connections.

As we travel back to the present, one may ask – how does all this relate to small-world networks? Well, these trade routes were essentially creating shortcuts across vast distances, much like how Watts and Strogatz's small-world model adds random connections to a regular network.

Each trader along these routes acted like a node in the network. While most connections were local (like trading with neighbouring cities), some long-distance connections (like the occasional trader traversing the entire Silk Road) created those crucial shortcuts that define small-world networks.

This structure allowed for the rapid spread of not just goods, but also ideas, technologies and even diseases (here's looking at you, bubonic plague!). It's why you can find Buddha statues with Greek-inspired designs in Afghanistan, or why chess – a game that originated in India – became popular in Persia and then Europe.

The efficiency of these trade networks in spreading information and innovations mirrors the high clustering and low average path length characteristics of small-world networks. Ideas could 'hop' across vast distances through these trade connections, much like how information spreads rapidly in modern small-world networks like the Internet.

Imagine Shefali, a savvy trader from Pataliputra (modern-day Patna), setting out on the Grand Trunk Road with a cartful of fine Magadhan silk and a pouch full of Mauryan silver coins. Her most prized possession? A few meticulously crafted iron tools from the famed blacksmiths of her hometown. It's 250 BCE, and she's heard whispers of strange new goods from lands beyond the Hindu Kush. As she navigates the bustling thoroughfare, she can't help but feel like a data packet in Oreo's olfactory network, zipping along neural pathways of commerce that span continents.

At a crowded marketplace in Taxila, Shefali trades some of her silk for a curious new spice – black pepper from the Malabar Coast. She also picks up strange tales of Yavana (Greek) philosophers and their radical ideas about atoms and the void. As she samples some unfamiliar fruit, she overhears a merchant from Bactria boasting about a new technique for smelting iron. Shefali grins, knowing the blacksmiths back home would scoff at this 'new' technique. After all, Indian metallurgists had been producing high-quality wootz steel for centuries, a material so advanced it would baffle European scientists two millennia later. Think Damascus blades and the rust-resistant Iron Pillar of Delhi.

During this journey, Shefali's network has expanded. She now has contacts stretching from the pepper farms of Kerala to the vineyards of Greece, and from the iron ore mines of Bastar to the sword-makers of Persia. She has become a living breathing 'random connection' in the great tapestry of ancient trade and technology transfer – much like that one auto driver in Bangalore who somehow knows a shortcut that isn't on Google Maps, and also how to fix your car's engine with just a rubber band and a coconut shell. Shefali's journey can be seen as part of a vast small-world network, where her iron tools inadvertently helped spread Indian metallurgical knowledge far and wide, influencing technologies ranging from Arabic pattern-welded swords to English Sheffield steel.

As Oreo drags me to the lane containing all the bakeries, let me take you from Shefali's story to mine. Close to my hometown is the sprawling old city of Varanasi. I can smell the freshly cooked puris when I think about it. I can remember standing on the banks of the Ganga watching the sacred waters flow by. At first glance, the river's movement seems chaotic, a tumultuous dance of currents and eddies. But look closer, and you'll start to see patterns emerge that would make any network theorist's heart skip a beat.

Water, you see, is a bit of a rebel. It doesn't always take the path of least resistance – sometimes it creates its own shortcuts. As the river flows, it forms what hydrologists call 'preferential flow paths'. These are like the popular kids in a high school social network – everyone wants to connect with them.

In a river, you might see a main channel where most of the water flows swiftly. This is like the 'highway' of the river network. But then you'll also spot smaller streams and rivulets branching off and reconnecting. These act like the random long-range connections in a small-world network, creating shortcuts that help the water navigate complex terrain more efficiently.

During floods, this network structure becomes even more pronounced. The floodwater doesn't just spill over evenly across the land. Instead, it finds and creates preferential paths, forming a complex network of flows that can transport water, sediment and, unfortunately, sometimes your neighbour's garden furniture, over long distances incredibly quickly.

As you look, the roads along the banks of the Ganga light up with electric lamps.

So, let's flip the switch and talk about electricity. If water is the rebel teenager of the natural world, electricity is its law-abiding, lightning-fast cousin. Surprisingly, though, it behaves in a very similar way when it comes to network patterns.

Picture the power grid of a city like Mumbai. You've got your big power stations, high-voltage transmission lines, local substations and, finally, the lines that bring electricity to your home. At first glance, it might seem like a straightforward, hierarchical system. But in reality, it's a beautiful example of a small-world network.

The main high-voltage lines act like the main channel of a river, carrying the bulk of the power. But the grid also has numerous interconnections and backup routes. These are like those small streams in a river system – they might not carry as much power, but they provide crucial shortcuts and alternative paths for the electricity to flow.

This small-world structure is why electricity doesn't have to travel all the way from the power plant through every substation in the city when you switch on your light. Instead, it finds the most efficient path to your home, often using these 'shortcut' connections.

It's also why the lights don't always go out across the

entire city when there's a power outage in one area. The small-world-network nature of the grid allows electricity to reroute quickly, finding alternative paths to keep as much of the network powered as possible.

Interestingly, both water and electricity networks face similar challenges. Just as a river might have to navigate around a mountain, electricity has to find its way around areas of high resistance. And just as a flood might overwhelm certain paths in a river network, a surge in electricity demand can overload certain parts of the power grid.

The similarities don't stop there. In both cases, engineers and nature use similar strategies to manage flow. In a river, you may see natural levees forming along preferred flow paths, reinforcing these channels. In an electrical grid, engineers may upgrade certain transmission lines that form key shortcuts in the network.

So, from the meandering path of a river to the unseen journey of the electrons powering your home, you're witnessing the principles of small-world networks in action. It's a beautiful reminder that the same mathematical principles can describe everything from your social media connections to the flow of water and electricity.

As our walk closes in towards our destination, I look at the hustle and bustle of the morning market waking up, and I am transported back to my childhood.

Picture this: Tiny me, all of 10 years old, thrown into the chaos of the old market of Allahabad (now Prayagraj), tasked with the impossible. My mission? To find a skein of marigold orange wool. My teacher – who clearly had no sympathy for Indian summers or my tendency to procrastinate – thought crafting flowers out of yarn was the educational pinnacle of fourth grade. Just like any other self-respecting child, I waited till the day before the summer holidays ended to start working on the project.

The market was a labyrinth of narrow lanes, each one seemingly more convoluted than the last. The baby pink walls of the shops glowed in the late afternoon light, their vibrant hue mocking my desperation. With every turn, I became increasingly certain I had stumbled into a parallel dimension where wool was rarer than silence in my household. It was the kind of sprawling chaos that felt like the opposite of order.

And yet, as I meandered aimlessly, I wasn't really alone. This market wasn't just a collection of shops; it also was a living, breathing network of connections – shopkeepers who knew your dad's cousin's best friend, florists who remembered your mom from a wedding years ago, strangers who somehow felt like distant relatives. Each person was a node in an intricate web of relationships, their paths crisscrossing in ways that defied logic but worked nonetheless.

Enter my mom: the ultimate connector in this small-world network. Unlike me, she didn't feel out of place and overwhelmed, though she was extremely annoyed at me for not having told her about the project until the last possible minute.

She grabbed my hand and dived into the market, navigating it with the expertise of someone who had long mastered its unwritten rules. The pink walls seemed to bow to her authority. I was the unassigned node, while she was the central node in a large network.

We didn't rely on luck; we relied on connections. Mom

spotted an old shopkeeper she knew from years ago, a man who had once sold her fabric for a sari. A quick conversation later (peppered with updates on mutual acquaintances, and an invite to the shopkeeper's daughter's wedding), he directed us to a tiny wool shop tucked into a corner I would never have found on my own. Indeed, I could've sworn it actually didn't exist five minutes ago. It was so small it could've doubled as a cupboard, but it held exactly what I needed. The marigold orange skein was plucked from a shelf that felt like a portal to a world where homework disasters were averted.

This wasn't just about wool; it was about how markets like these operate. Every shopkeeper seemed to know someone who could help. When one node in the network didn't have what you needed, they'd point you to another, their connections spanning the market like an invisible map. What felt overwhelming to me as a 10-year-old wasn't chaos at all – it was a beautifully interconnected system, and while I understood it intuitively then, I wouldn't acquire the language to describe it until much later.

This, my friends, is the essence of a small-world network: a system in which individuals (or nodes) are connected through relatively few intermediaries. The market might seem vast and disorganized, but thanks to these overlapping connections, solutions were just a few conversations away. It's the same phenomenon that made Milgram's six degrees of separation work. In this case, the network wasn't just theoretical – it was woven into the market's very fabric.

On the way home, clutching my precious wool, I realized how much of this experience relied on connections I hadn't even known existed. My mom's knowledge of the market, her relationship with the shopkeepers and their ability to direct us to the right place were all examples of how small-world networks function in real life. The market may seem sprawling, but in reality it's a tightly knit web in which solutions are never far – if you know the right people.

As Oreo stops to thoroughly investigate the tyre of a Maruti Swift (spending so much time there you'd think he was trying to decode the Da Vinci Code), I realize his exploration is more than just idle curiosity. Each sniff adds to his mental database, updating his understanding of the neighbourhood's happenings. Who's been here? When? Were they excited? Scared? He wags his tail to communicate them all to me but, unfortunately, his small-world networks and mine aren't connected that well yet.

As we round the corner back onto my street, it hits me – Oreo's seemingly random path is anything but. Every sniff, every detour, every inexplicable pause at a lamppost is part of a carefully optimized route through his own small-world network of smells, social interactions and strategic decisions. In his way, Oreo is as skilled at navigating his world as my mum was at navigating the bustling markets of Allahabad – or as ancient traders were at traversing vast trade routes, using the nodes they were connected with and just the right shortcuts to get things done.

And, just like that, we're back home. Oreo's 'inbox' is fully checked, his mental map is updated and his diplomatic missions accomplished. His walk, much like the networks we all rely on, shows how a world that seems vast and chaotic is actually tightly connected in surprising ways.

So, remember – the next time you're marvelling at your

grand aunt's story about how the president is basically a family member through your uncle's cousin's neighbour's aunt's brother-in-law's friend, you're actually witnessing the mathematical magic of small-world networks in action. In this vast, complex world, it seems everything is more connected than we might think – and often in the most elegant and efficient ways possible.

It's a small world, after all!

4

Beauty and the Bloom

After the walk, I decide to sit down and paint. I take out my watercolours and stick some high GSM watercolour paper onto a piece of wood, covering the whole thing with broad strokes of plain water, which is the best way to spread water-based paint evenly. Oreo, correctly understanding my actions to mean that I am about to paint, takes up his post on the bed where a patch of sunlight is peeking through the window, getting comfortable.

Oddly, though he often bullies me into giving him pets and treats and zooms around the flat knocking down stuff when I'm not painting, he has never once spilled my paint water or stepped on my paintings, even when these are left untended on the floor to dry. Maybe that's one of the ways in which he shows he really loves me. He rolls into a ball, perfectly fitting himself into the warm light beam.

As I wait for the paper to dry, I set up my watercolours by the window, trying to capture the way the morning light plays on my drumstick tree's flowers. The white petals have a luminous quality that makes them almost impossible to paint; how do you show light on white? My amateur attempts at botanical illustration are a far cry from the precise watercolours of nineteenth-century botanical artists, but simply trying to paint the flowers makes me look closer at them and notice more detail. There's an asymmetry to these blooms that I never saw before I tried to put them on paper.

It was a while ago that I got drawn into painting flowers. It began as an effort to capture the ephemeral beauty of flowers around my college campus at the National Centre for Biological Sciences in Bangalore. Pictures could never capture it fully, but my walks through the campus showed me a living gallery of flowering trees. The landscaping there was deliberately wild, almost rebellious against the precise science happening within the buildings. I remember walking between experiments, past beds of jackfruit trees competing with mystery yellow flowers, while a string of trees carpeted the earth every morning with their pink powder-puff blooms.

I also started trying to capture the lovely birds of paradise flowers, which mimic the birds of paradise in flight at the entrance of the institute, or the gulmohars above. The wonderful blooms around the campus made the fatigue caused by squinting at code that didn't work vanish, if only for a moment. (Incidentally, my code would refuse to work until the moment I surrendered and asked for help from someone; Murphy's law obviously meant that in front of that other person, it would suddenly start working.)

But it was another of my haunts, the Gandhi Krishi Vigyana Kendra (GKVK) or the University of Agricultural Sciences, that held the real surprise. Hidden behind the agricultural university's practical exterior was a wonderland of botanical diversity. I'll never forget stumbling upon a field of sunflowers late one evening – hundreds of golden heads all turned westward, tracking the setting sun. It was like walking into a van Gogh painting, but one where the flowers moved – as indeed they seem to even in van Gogh's painting.

Now, I dip my brush in water and try to mix just the right shade of cream for my drumstick blossoms. This requires a precise mix of yellow, pink, grey and surprisingly, blue. I think about how those trees have been silent witnesses to generations of scientists.

Next to my current focus, aka the drumstick tree, is its neighbour – the lovely pomegranate tree. It is having its own celebration, its bright orange-red flowers looking like tiny, crumpled silk lanterns. These vermillion blooms, shaped like miniature vases, are actually why the Chinese call it the 'flower vase fruit'. Each blossom looks like it's been crafted by a master origami artist – first folded precisely, then crumpled just enough to create those characteristic wrinkles in its waxy petals.

The flowers tell stories older than any human civilization. Just like they do in my paintings, these flowers make their mark across civilizations, showing up in stories and myths.

In ancient Persia where the pomegranate tree originated, these blossoms were seen as drops of the blood of Esfandiyar,[1] a mythical Iranian hero whose story is older than written history. The pomegranate, with its many seeds, symbolized abundance and life, but the very act of breaking open the fruit also implied decay and death – a duality that perfectly mirrors Esfandiyar's own fate, for his life took a turn for the

---

1 An invincible hero from the *Shahnameh*, the Iranian Book of Kings.

worse after he broke open a pomegranate. Just as the seeds of the pomegranate are contained within a tough skin, life and death are entwined in his destiny.

Greeks wove the pomegranate into the tale of Persephone, the unwilling Goddess of the Underworld. When she was abducted by Hades, the God of the Underworld who was in unrequited love with her, the only thing she ate was a pomegranate, and that doomed her to spend some time 'downstairs', only returning to the world upstairs to spend spring and summer with her mother Demeter, the Goddess of agriculture. The legend claims that the number of red seeds of the fruit matches the number of winter days she must spend in the Underworld.

In Judaism, the fruit is said to contain 613 seeds, matching the number of commandments in the Torah, though any botanist with enough patience to count will tell you the number varies quite a bit.

The flowers themselves are botanical overachievers. While most fruit trees need different varieties planted nearby for pollination, pomegranate flowers contain both male and female parts arranged in two distinct whorls – some perfect flowers[2] up top that will form fruit, and some male-only flowers below that exist purely for pollen production.

As the flowers mature, you can actually predict which ones will become fruit. The ones with a bulbous base, looking like they're wearing crinolines[3] under their red dresses, are the

---

2 Aka bisexual flowers – having both male and female reproductive parts.

3 You know the things that made women look like they were hiding an umbrella inside their dresses?

ones that will transform into those leather-skinned fruits filled with jewel-like seeds. The others, the skinny ones, will drop after they've shared their pollen, their job done.

The British may have brought their mahogany and gulmohar, but these trees have their own ancient stories, older than colonial gardens and city plans. The pomegranate has been grown in Indian courtyards since the time of Harappa, its flowers gracing gardens from Kashmir to Kanyakumari. When you look at these blooms, you're seeing the same shapes that inspired the border patterns on Mughal textiles, the same colours that dyers tried to capture in silk, the same flowers that have been painted in miniatures and carved into temple walls for thousands of years.

I decide to add the red flowers to my painting. However, I don't pick red. I start by painting the undertones, the greys and the darks, on the canvas.

As I do, I breathe in. The morning air carries that distinct Bangalore fragrance – a mix of exhaust fumes and flower petals, depending on which way the wind blows. Today, it's bringing me whispers of jacaranda blooms from the next street, mingling with the sweet scent from my drumstick tree's flowers. This is a good day. Sometimes the breeze brings me the stench of overflowing gutters instead; it's a gamble that I take with every deep breath I take. The purple petals of jacaranda have been carpeting the sidewalks lately, making every morning feel like a scene from a romantic movie.

'Did you know,' I tell a bulbul who's flown in and is now pecking at these fallen flowers, 'that none of these purple beauties are actually from here?' The bird, predictably, is more interested in finding breakfast than in botanical history, but the story of how Bangalore became a city that blooms all year round is worth telling, even to an uninterested audience. Though uninterested, the bird seems to find solace in the pomegranate tree, and I imagine it is ready to listen to my story.

As I load my paintbrush with more colour and focus on the painting, let me tell you the story of what happened. It started, as many things in old Bangalore did, with the British and their peculiar obsession with making everything in their colonies look like 'home'.[4] They brought seeds and saplings from across their empire, turning Bangalore into an experimental garden. The jacarandas came from Brazil, the copper pods from Southeast Asia, the African tulips (obviously) from Africa (unlike French toast or French fries, neither of which are from France). It was like hosting an international flower show that never ended.

A sudden gust of wind showers me with tiny white neem flowers. Even after all these centuries, the neem remains stubbornly local in its habits – flowering exactly when the Karnataka summer begins to show its teeth. Its bittersweet fragrance is nature's way of announcing that it's time to stock up on tender mangoes for pickle-making. Speaking of mangoes, their trees are putting on their own subtle show right now – delicate cream-coloured panicles that promise sweet rewards in a few months. It's fascinating how these unassuming flowers transform into the king of fruits, each variety from Alphonso to Raspuri starting its journey as these

---

[4] Even though they turned their museums at home into exotic destinations by stealing artefacts from all around the world.

modest blooms.

A little further up from here, near Cubbon Road, there's a canopy of rain trees. Their branches reach across the street like old friends holding hands, creating patches of shade that make Bangalore summers bearable. If you take a walk there and look up, you'll catch sight of their delicate pink powder-puff flowers, scattered among the leaves like tiny fireworks frozen in time. These gentle giants are perfectly named – when their tiny leaflets fold up before rain, they seem to whisper secrets about upcoming showers to those who know how to listen. These giants weren't part of the original plan either. Some colonial botanists thought they'd make good shade trees for soldiers, and now here they are, centuries later, still faithfully providing shelter to rushed commuters and tired street vendors.

Bangalore's high altitude and temperate climate made it the perfect experimental ground for the flowering experiments. The British saw in its gentle weather and rich soil an opportunity to create a botanical bridge between continents. Tabebuias from South America, jacarandas from Brazil and gulmohars from Madagascar found themselves growing alongside India's native sampige and akasha mallige. The colonial botanists, armed with their specimen books and botanical illustrations, meticulously documented how these foreign species adapted to their new home.

When Gustav Hermann Krumbiegel arrived in Bangalore circa 1891, he inherited this living collection of global flora. What set the German botanist apart was his vision. He transformed these almost random botanical experiments into a huge artistic masterpiece using the entire city as his canvas.

Working with Indian horticulturists who brought generations of local knowledge about seasonal rhythms and growing conditions, he orchestrated what would become Bangalore's famous flowering calendar.

The native trees already had their roles – the sampige had been perfuming Indian gardens for centuries, and the silk-cotton tree had been putting on its dramatic show long before any botanist thought to document it.

What Krumbiegel and his team did was create a harmonious dialogue between these indigenous performers and their non-native counterparts. They noticed how the pink tabebuias could extend the spring show just as the native blossoms were fading, and how the jacarandas' purple canopy provided a perfect backdrop for the golden show put up by the Indian laburnum.

This wasn't just botanical matchmaking – it was also cultural fusion at its finest. The British had unknowingly set the stage by bringing trees from their various colonies, but it was the combination of European botanical science, Indian horticultural wisdom and Bangalore's lovely climate that turned it into art. The umbrella trees from the Pacific Islands learnt to dance with the pride of India trees, while South American plumeria found itself in fragrant conversation with Arabian jasmine. Think of it as the most advanced version of an exchange program.

These trees, which had evolved continents apart, created new seasonal rhythms together in Bangalore. The jarul trees, native to South Asia, discovered that they could share the spotlight with African flowering species. Even the mahogany trees, primarily valued for their timber in their native

Americas, found a new identity in Bangalore as flowering performers.

What remains now is one of the greatest flower shows in the world. Every month brings a new dominant colour to the city's streets and parks. It's like Bangalore can't decide which season it likes best, so it celebrates all of them at once.

In January, when *Tabebuia rosea* launches the year's celebrations, entire avenues turn into corridors of pink. Their pink blooms glow against the winter sky like perfectly positioned spotlights, as if nature hired its own lighting director. March belongs to the purple-robed royalty of jacarandas, but April ... Oh, April is when our story reaches its crescendo. It's like someone calls 'action!' and every tree decides to give a show-stopping performance simultaneously, as if competing in an ancient beauty pageant judged by the gods themselves. The copper pod trees steal scenes with their dramatic yellow cascades, creating such thick carpets of flowers that the sidewalks disappear beneath nature's own red-carpet treatment. The Indian laburnum, aka amaltas, joins this spring symphony right on cue, unleashing its golden display like strings of marigolds hung for a celebration. In May, the gulmohar trees burst into bloom, their flame-red flowers setting the sky ablaze like a thousand diyas lit for a festival.

It is beautiful indeed, like a dance across months. The tabebuias execute their perfectly timed colour guard performance – first come the pink flowers (quite often known as Bangalore's cherry blossoms, though the name is a misnomer) known as *Tabebuia rosea* in January, then passing the spotlight to its yellow cousins glowing like morning sunshine, which announces spring's arrival with their delicate yellow trumpets. Each flower knows its cue perfectly, like dancers in a carefully choreographed Bharatanatyam performance.

Just before the pink tabebuias can take their final bow, the silk-cotton tree (*Bombax ceiba*) makes its grand entry from March to May, like Rajinikanth in a mass melee scene. Its fiery red and orange flowers command attention with the confidence of a seasoned superstar – bold, dramatic and impossible to ignore. These aren't just flowers; they're also nature's own standing ovation to herself.

The sampige (*Michelia champaca*) follows this act from June through September, releasing a fragrance sweeter than fresh jalebis on a monsoon evening. Their golden-yellow flowers perfume the air like nature's finest attar, with some rebel blooms making surprise appearances all year round, because who said beauty needs a schedule?

As this heavenly scent fills the air, akasha mallige (tree jasmine) weaves its magic from April through September (and sometimes beyond). These tiny white flowers perform like classical musicians – their subtle fragrance creates ragas in the air that only the heart can hear. The jarul trees (*Lagerstroemia speciosa*) wait patiently before bursting into a stunning display of purple-pink flowers during July through September. Though often mistaken for jacarandas, their blooms hold fast to their branches, determined to keep the show going rather than carpeting the ground.

Nearby, the umbrella trees (*Thespesia populnea*) add their own nuanced rendition – yellow flowers that mature into burgundy like a perfectly aged wine. The pride of India trees enhance the composition with their lilac blooms, while even the usually serious mahogany trees join in with their tiny white flowers. These gentle giants, despite their deeply furrowed bark and massive canopies, seem almost shy about their delicate blooms – like action heroes revealing their poetic side.

And then there are the perennials – the star performers who keep dazzling all year round, no matter the season. Think of the bougainvillea. It's the city's most enthusiastic costume designer – no wall or fence is too challenging for this artistic draper. Year after year, it transforms ordinary compounds into extraordinary spectacles with its paper-thin bracts in every shade, from the deepest magenta to sunrise orange. Some say it's not a true flower putting on the show – those bright colours actually come from modified leaves – but isn't that just like a brilliant costume designer, creating magic from the unexpected?

Then there are the rain trees (*Samanea saman*) – another of our gentle giants, the steady performers who believe in subtle but consistent showmanship. Their soft pink powder-puff flowers float down like botanical confetti throughout the year. These trees don't just perform – they create entire ecosystems, their broad canopies offering a stage for birds, bees and butterflies to enact their own daily dramas.

There's also the temple tree (*Plumeria rubra*), blooming away like it's got a lifetime contract with nature's production house. Its fragrant white and yellow (or sometimes, pink and yellow) flowers appear with such reliable regularity that you could practically set your calendar by them. Their sweet perfume is so consistent that early morning walkers plan their routes to pass under these botanical perfumers, each flower a testament to the idea that some of the best performances are the ones you can count on seeing every single day.

This magnificent production runs year after year, each season bringing its own colour, fragrance and magic. It's nature's own streaming series, but better – no subscription needed, no ads to skip, just pure, uninterrupted beauty playing out in the canopies above our heads. Just look up, and let the show begin!

'It's chaos,' my botanist friend once said, 'beautiful chaos.' She's right. In nature's grand design, trees are supposed to flower in sync with their species, creating reliable seasonal patterns – but something went sideways in Bangalore, and it's wonderful. The moderate climate confused the internal calendars of trees from different continents. Instead of sticking to their native schedules, they adapted, creating a year-round blooming cycle that exists nowhere else.

I change my brush, picking up the finer one to add details. I open the window further to let the air and the smell of the blossoms in.

The old banyan at the corner has been here long enough to remember when the street was a mud path. Its aerial roots have grown into extra trunks, creating a small forest from a single tree. Unlike its showier neighbours, the banyan flowers discreetly, its tiny blooms hidden inside figs that feed entire colonies of birds and insects.

Looking up at all these blooming trees, you might think flowers are nature's greatest success story. But here's the strange thing – flowering is actually a terribly inefficient way to reproduce. It's like hosting an elaborate wedding when you could just send the marriage registration papers by post.

These trees invest enormous energy in creating delicate petals, producing sweet nectar and synthesizing fragrant compounds, all for a small chance at reproduction.

So why go through all the trouble? Why not just keep things simple, like mosses and ferns, who've been happily getting along by releasing spores – no bells, no whistles, just plain ol' 'send it and forget it'? They make spore after spore, each one genetically identical and ready to grow at the drop of a hat. It's easy, it's efficient and it's been working for millions of years.

The answer, like everything else in biology – and life – is evolution[5]. Flowers are the ultimate gamble. They have a flashy, high-risk strategy that ultimately pays off in spades.

Early plants like mosses and ferns relied on water and wind to get their spores around. They were passive, just sending out spores and crossing their fingers. But flowers? Flowers stepped up their game: They recruited animals to do the hard work. Instead of relying on water or wind, flowers decided to use bees, birds, bats and us humans to spread their pollen. Pollen is simply the tiny particles from a flower that carry the male reproductive cells of the plant. These little particles need to be transferred to the female part of another flower for the plant to reproduce. That's where pollinators come in. Pollinators are creatures like bees and butterflies that move pollen from one flower to another, often without even realizing they're doing it. As they visit flowers to collect nectar (a sugary liquid that they use for food), they end up carrying pollen from flower to flower, helping plants reproduce in the process.

---

5 There's even a famous quote about this.

It's like plants found their own delivery service, and it turned out to be way more reliable. Now they weren't just hoping for the wind to blow in their favour; they were *targeting* their pollinators. It was a brilliant move – suddenly, they weren't just reproducing; they were ensuring that their genes got around in the best way possible. And more genetic diversity (variety in the plants' offspring) meant better chances of survival.

But here's the kicker: Flowers didn't just offer pollen and pollen alone. No, they went the extra mile and offered rewards. Nectar. Sweet, sweet nectar. And let's face it, if you're a pollinator, free snacks are hard to turn down. In return for all that deliciousness, the pollinators did the hard work of ferrying pollen from one flower to another, making the whole process far more reliable. It wasn't just a one-way street – flowers got better at being attractive, and pollinators got better at their jobs. Over some 150 million odd years, this back-and-forth has led to a mind-boggling variety of flowers, each with its own twist on the game.

One of the most fascinating results of this was the way flowers and their pollinators became so perfectly matched. Take, for example, a puzzle that Charles Darwin, the scientist who first proposed the theory of natural selection, stumbled upon in 1862. He was studying an orchid from Madagascar that had a bizarre feature – a very long, narrow tube, almost as long as your forearm, with just a tiny drop of sweet nectar at the bottom. Picture trying to drink from a straw that's longer than any straw you've ever seen; that's what any creature would have to contend with to reach the nectar.

As Darwin examined this strange flower (now called

Darwin's orchid), he made a guess that sounded pretty far-fetched at the time. He proposed that there must be a moth out there with a tongue (technically known as proboscis) long enough to reach that nectar – a tongue that could stretch out to the length of a pencil. No such animal was known at the time, and his prediction drew scepticism. A moth with a tongue that long? It seemed like something straight out of a fairy tale.

But Darwin stuck to his guns. He figured that if the flower existed this way, there must be a reason. It's like finding a lock and knowing there must be a key somewhere that fits it perfectly. The flower and its pollinator, he believed, would have grown and changed together over time, like dance partners learning each other's moves.

It took almost 150 years, but in 2004, scientists finally found exactly what Darwin had predicted – a moth with an incredibly long tongue that could reach the bottom of that orchid's tube. They named it *Xanthopan morganii praedicta* – adding that last part 'praedicta' (meaning 'predicted') as a tip of the hat to Darwin's remarkable foresight.

The Victorians, those masters of hidden meanings and repressed desires, would have appreciated such floral mysteries. In an era when direct expression of emotion was considered vulgar, they developed floriography – the language of flowers – into an art form so complex it required dictionaries to decode. It wasn't just a quaint hobby; it was a sophisticated underground communication system, as intricate as modern encryption methods.

Lady Mary Wortley Montagu had imported the idea from the Ottoman Empire, where 'selam' (the Turkish flower language) allowed the women in the harems to communicate in secret. In her *Turkish Embassy Letters* (written around 1717), she described the 'selam' system used in the Ottoman Empire. She famously wrote, 'There is no colour, no flower, no weed, no fruit, herb, pebble, or feather, that has not a verse belonging to it; and you may quarrel, reproach, or send letters of passion, friendship, or civility, or even of news, without even inking your fingers.' The Victorians took this concept and ran with it, developing it into an obsession that spawned dozens of flower dictionaries and guided everything from courtship to condolence.

A proper Victorian lady might receive a bouquet that read like a love letter, each flower carefully chosen for its hidden meaning. Tuberose, with its intoxicating night fragrance, spoke of dangerous pleasures. Jonquil's bright yellow trumpets declared desire, while gardenias whispered of secret love. A spray of lily-of-the-valley might have suggested a promise to 'return to happiness'. The addition of violets – those humble purple blooms – carried special significance in certain circles, for they had long been associated with safeguarded love between women, dating back to Sappho's ancient poems.

The messages weren't always romantic. A bouquet could wound as effectively as any cutting remark. Yellow carnations screamed disdain, striped roses spelled rejection and orange lilies declared hatred. Some arrangements were downright threatening – a perfectly arranged bouquet of beautiful but poisonous flowers like foxglove or rhododendron carried warnings of insincerity or danger. Aconite, also known as monkshood, with its deep purple hooded flowers, was particularly dreaded – it meant 'beware, a deadly foe is near'.

And then there were the 'unnatural flowers'. In the glittering world of late Victorian London, more specifically 1892, Oscar Wilde sparked a peculiar fashion that later contributed to his trial and prison sentence. He asked his young friends and admirers to wear green carnations in their buttonholes to the opening night of his comedy *Lady Windermere's Fan*. Natural carnations, of course, don't come in green – these flowers had to be artificially dyed by leaving white carnations in green-tinted water.

Their very artificiality was part of the point, a celebration of the unnatural, and a secret badge of recognition for members of the LGBT community. These artificial blooms – these deliberately cultivated unnaturally coloured flowers – became a symbol of what Victorian society called 'unnatural' affections. In an era when homosexuality was not only socially taboo but illegal, the green carnation served as a kind of secret handshake, a way for men who loved men to recognize each other in London streets and theatre lobbies.

The green carnation became a subtle code among Wilde's circle. When Robert Hichens wrote a roman à clef (a thinly fictionalized novel) about Wilde titled *The Green Carnation* in 1894, everyone in London society understood its implications.

The symbolism would later be used against Wilde during his trial. The prosecutor pointedly asked about the meaning of the green carnation, trying to establish it as evidence of 'unnatural' behaviour. Wilde, with his characteristic wit, deflected: 'Nothing whatever but a fashion.' But the damage was done. The green carnation had become too well-known as a symbol, its meaning too clear to those who wished to condemn him.

The flowers that had once served as playful badges of defiance against Victorian conventionality were transformed into exhibits in a criminal case. They joined other 'evidence' – the posed photographs, the yellow books, the silver-gilt cigarette cases – in the prosecution's attempt to paint Wilde as a corrupting influence on young men. The very artificiality that made the green carnation so perfect as a symbol of aesthetic rebellion made it equally perfect as a symbol of what society deemed perverse.

After Wilde's conviction and imprisonment, the green carnation largely disappeared from London's buttonholes. The code had been broken, the secret exposed. But its legacy lived on in other forms of floral communication. Those who needed to hide their true feelings found new flowers, new arrangements and new ways to use the language of blooms to say what could not be said in words.

For example, a flower turned to the right could mean 'yes' while one turned to the left meant 'no', a convention that dating apps like Tinder seems to have adopted. Upright flowers conveyed positive thoughts, while inverted blooms suggested the opposite. Pansies, whose name comes from the French *pensée* (thought), could mean 'think of me' when upright, or 'forget me' when inverted.

The intricacy of these codes served those whose love dared not speak its name well. Violets paired with rue might suggest regret for a secret love, while ivy entwined with myrtle could promise eternal fidelity outside society's bounds. Dried white roses – which meant 'death is preferable to loss of virtue' in conventional bouquets – took on different meanings in private exchanges between those who had to hide their true nature.

Even the ribbon carried meaning. Tied to the left, it meant the flower's symbolism applied to the giver; to the right, the recipient. The type of knot, the length of ribbon, even the shade of colour could modify the message. A violet ribbon might transform an innocent bouquet into a declaration of hidden passion, while black ribbon could turn the same flowers into an expression of shared sorrow at having to maintain public facades.

It was social media before electricity, more layered than any modern emoji. Every petal, position and combination could carry multiple meanings, depending on who was sending and who was receiving the flowers. In private conservatories and secret gardens, these floral codes allowed the expression of feelings that Victorian society refused to acknowledge. The same bouquet might have carried one meaning when presented at a public gathering, and an entirely different one when passed discreetly in a darkened parlour or a quiet corner of a garden.

The complexity of these codes meant that many messages were lost to history, pressed flowers in old books whose secrets died with their givers and recipients. Yet some survived, preserved in diaries and letters, revealing how those forced to live in society's shadows found ways to speak their truth through nature's own alphabet. When we talk about Victorian flower language today, we often focus on the romantic aspects – the courtship and scandal – but miss how these codes served as lifelines for those who had no other way to express their authentic selves. In an era of rigid social control, flowers provided a voice for the voiceless, a way to say what could not be said aloud.

What's amusing, though, is how these elaborate human codes pale in comparison to the actual messages flowers are sending. I remember a rain tree above me once casually dropping a cluster of its pink powder-puff blooms, covering me in a sudden shower of flowers, and I can't help but smile at nature's extravagance. These flowers are speaking in a language so sophisticated it makes Victorian floriography look like baby talk.

What we see as pretty pink or purple petals are actually more like a giant billboards with messages written in invisible ink – invisible to us, that is, but crystal clear to bees and butterflies. Think of it like this: imagine if traffic signals had secret messages that only taxi drivers could see, telling them exactly where to find passengers. That's what flowers are doing, but instead of passengers, they're advertising nectar and pollen.

For instance, the copper pod's yellow cascade isn't just for show. Those bright petals are broadcasting in ultraviolet wavelengths, creating patterns that we'd need special cameras to see. These UV patterns, invisible to human eyes, often form bull's-eye patterns leading to the flower's centre, like runway lights guiding their insect pilots. Some flowers even change their UV signals after pollination, switching off their 'landing lights' to tell bees 'nothing to see here, move along'.

While these flowers are putting on a dazzling display for their insect visitors, they also happen to be an example of what botanists call 'perfect flowers'. This is a very appropriate term for something quite lovely. 'Perfect flowers' or 'bisexual flowers' house both male and female parts in the same bloom. It's an efficient, self-contained setup for reproduction, designed to make sure the work of creating new plants goes off without a hitch. Perfectly efficient in a world that also values beauty.

It's quite practical when you think about it. When you're a tree and can't move around to meet other trees, it makes sense to have all your important reproductive parts in one place.

It's a reminder that nature's solutions often transcend our human categories and concerns. While Victorian society was constructing elaborate social rules around gender and reproduction, their gardens were filled with perfect flowers that quietly demonstrated a more fluid and practical approach to the business of survival.

Yet, despite there being perfect flowers, most trees avoid self-pollination like Victorians avoided scandal. They've developed intricate timing mechanisms where male and female parts mature at different times, ensuring self-pollination is prevented within the same flower. It's like having a built-in chaperone at a Victorian dance – everything appears to be prim and proper while still allowing for a bit of necessary mingling.

The drumstick tree greets me with another shower of flowers as I add the final touches to my painting. I smile, thinking to myself …

The British wanted to create a little England in Bangalore. Instead, they accidentally helped create something far more interesting – a city that celebrates spring all year round, where trees from different continents learnt to dance to the same seasonal rhythm.

As I brush the flowers from my hair, I ponder how these trees represent one of evolution's most fascinating gambles. About 130 million years ago, some ambitious plants decided to try something new – instead of simple spores or cones, they invented flowers. It was like switching from a reliable postal service to an elaborate system of skywriting and fireworks to send messages. By all logical measures, it should have failed spectacularly. Yet, here we are, under a canopy of blooming trees that have turned their reproductive strategy into an art form. These flowers are living proof that sometimes, the most beautiful things in nature arise not from efficiency but from delightful extravagance.

Oreo chooses this moment to wake up from his nap and come to me for pets; for him, of course there's no such thing as excess. I pause, pet him and take a look at the painting. It has elements I did not originally envision, but now I can't imagine the painting without them. In the same way, the trees around Bangalore may not originally have been part of the city but, if anything, they tell a story of adaptation – of how a place becomes home not because you were born there, but because you learnt to bloom there.

It's a statement true not just in the trees around us, but also in Bangalore's flower markets and temples, where flowers are more than just plants; they're a celebration of life, colour and sheer excess. If you walk through the bustling market aisles, you're surrounded by so much beauty it almost feels like the city itself is draped in a million different hues. The sellers arrange their blooms with love, creating a riot of colour and fragrance that invites admiration. It's a place where flowers – wild, manicured or cultivated – are allowed to be as extravagant as they want to be.

And here I am. As the painting comes together, I stretch, looking at the effort I made – not for efficiency, nor for profit, but simply for my delight. I think the flowers would approve.

5

Small Elephants and Big Dreams

It's that time of the day and week! The dreaded Sunday lunchtime. Do I order food from outside? Or do I cook? If I decide to cook, what do I make? I stroll into the kitchen, followed by an eager dog who hopes to receive some treats. Dog parents will understand: Oreo's treat supply is overflowing; my meal options, however, are close to non-existent. The irony isn't lost on me.

Let's see what I have – two potatoes and a tomato. Should I order more ingredients? No. I shall take up this challenge and cook something.

As I stare at the scarcity of ingredients to cook for myself, my mind drifts to tiny elephants – including the ones that were more or less the size of dogs. Wanna hear that story? It's a story of resilience, of adapting and of doing more with less. If you think about it, my current situation of having to cook a whole meal for myself is practically the same, right?

Don't answer that. Rather, let's start by addressing the elephant in the room – the room in question being my brain.

Elephants, the majestic giants of Asia and Africa, aren't just part of their ecosystems – they're the overachieving architects. These 5,000 kg landscapers reshape entire environments like they're auditioning for *Extreme Makeover: Habitat Edition*.

> Contestant 1 – Elephant A: 'I believe in zigzag paths. They're so aesthetic!'
> Contestant 2 – Elephant B: 'Um. I believe you should move as fast as possible and create a big path!'

They break branches, create clearings, dig waterholes and carve highways through dense vegetation that other animals happily freeload on. Even their dung isn't just dung – it's also a seed delivery system, planting forests one plop at a time. And with each elephant devouring up to 300 kg of vegetation daily, they're less 'gentle giants' and more 'hungry bulldozers' with an eco-friendly twist.

There's a reason why elephants feature so often in stories, in our mythologies, beyond simply being part of our existence.

What happens when you take these living compactors and trap them on a tiny island with limited resources? Do they stay large and in charge? Nope! They shrink, of course! Because nothing says 'adaptability' like becoming a pocket-sized version of yourself.

Okay, let's pause for a moment here and clarify something.

Elephants on an island don't exactly hold a group meeting and decide, 'Alright, gang, this island's too small for us. Let's all shrink a bit, yeah?' It's not like that. They have zero say in the matter. What actually happens is way sneakier – and, honestly, a little unfair.

Here's the deal: When food is scarce, being big and hungry is a liability. Smaller elephants, which need less food and can scurry over rocky terrain without breaking a sweat, end up surviving better. And since they're alive and kicking, they get to do what nature insists on – make more little elephants. Over generations, their genes dominate and, before you know it, the whole population downsizes, almost as though they were involuntary long-nosed fans of Marie Kondo.

So, it's not like the elephants are consciously choosing to go fun-sized. It's nature giving them a gentle but persistent nudge, saying, 'Smaller is better for this island, trust me.' And evolution? It's just a numbers game – tiny elephants win the survival lottery in scarcity (they need lesser to survive), and eventually, you've got a whole crew of adorable mini-trunkers wandering around. Evolution: 1, big elephants: 0.

Seems strange? You're not alone. The idea of tiny elephants was once considered so absurd that when archaeologists first stumbled upon the teeny-tiny skulls of these elephants in Mediterranean caves in the nineteenth century, they didn't think, 'Oh, cute! Mini-elephants!' No, they thought they'd uncovered evidence of the existence of cyclops – the one-eyed giant from the Greek myths. Yes, cyclops. Why? Because these skulls had a massive nasal cavity that, when viewed from a certain angle, looked suspiciously like a single eye socket. Honestly, it's not that absurd. Who in their right mind would expect to find elephants the size of large dogs?

The truth, of course, turned out to be less mythical but way cooler. In 1867, palaeontologist Hugh Falconer took a closer look and gave the first proper description of these pint-sized pachyderms[1] from Malta. It wasn't until the late twentieth

---

[1] Pachyderm – 'Thick skinned' – a term used to describe elephants and sometimes hippos and rhinos as well.

century, though, that scientists began to unravel the mystery of how and why these elephants shrank.

Spoiler: It's all about resource constraints. These little elephants weren't just nature's fun-sized experiment – they were proof of how evolution works when the buffet closes early.

The story hidden in their bones wasn't just a curiosity; it turned into a game changer for understanding evolution. It showed us that when resources dry up, even the largest creatures learn to downsize. Nature, as it turns out, has a thing for efficiency, and a sense of humour as well.

I decide to put the potatoes on to boil, while I look around my pantry for other ingredients I can use.

Back to the story – elephant miniaturization. How and why did it happen?

Imagine a time when vast, sprawling continents were connected by land bridges, and massive elephants roamed freely, grazing on abundant vegetation. These towering creatures were kings of their domain, their size a shield against predators and a tool for accessing food no other animal could reach.

But the world was changing. Glaciers melted, seas rose and land bridges disappeared beneath the waves, leaving behind scattered islands where some of these elephants found themselves stranded. Once mighty wanderers, they were now confined to pockets of land like Crete, Sicily and Malta, isolated from the mainland, from predators and from the lush resources they had once relied on.

At first, their size seemed a blessing. They could still forage, albeit with more effort, and with no lions or hyenas to fear, they seemed secure. But as time passed, the reality of island life set in. Food was scarce; the dense forests and expansive plains they once roamed were now replaced by rocky terrain and sparse vegetation. The elephant's massive frame, once an asset, became a liability.

It started slowly. In each generation, smaller individuals had a slight edge – they needed less food to survive and could move more easily over uneven ground. These smaller elephants reproduced more successfully, passing on their compact size to the next generation. Over thousands of years, the towering giants of the mainland gave way to something new: diminutive elephants no taller than a human, yet perfectly adapted to their island homes.

This transformation was no mere downsizing. This process led to changes in their entire physiology. Their bones thickened to support their smaller frames, their metabolisms became more efficient and their brains, though smaller in absolute size, grew proportionally larger compared to their bodies. These were not scaled-down replicas; they were entirely reimagined creatures.

Meanwhile, their cousins on the mainland grew no smaller. There, the evolutionary pressures were different: Size still meant strength and resources were plentiful. But on these islands, the elephants' smaller stature was their salvation. With every passing generation, evolution honed their efficiency, teaching them to do more with less.

Today, we can visit the Ghar Dalam cave in Malta, where layers of fossils tell the story of these prehistoric miniature pachyderms. The remains show a clear progression, each successive layer revealing smaller and smaller elephants, a testament to evolution's response to island isolation. Modern dating techniques have revealed that this downsizing happened relatively quickly in evolutionary terms, perhaps over just a few thousand generations – as opposed to millions of generations, which is the typical timeline for large mammals. ('Dating' here refers to techniques to find out how old something or someone is. So, actually, not that far removed from modern dating à la your favourite dating app.)

There's also today's Borneo pygmy elephants, which, while not as dramatically shrunken as their ancient Mediterranean cousins, also show how isolation and limited resources can drive size reduction. These living examples of evolutionary miniaturization provide a window into how nature solves the problem of resource scarcity.

I have still not started making my lunch, but I have found a small pack of potato chips. Obviously, I open up the pack and start munching while I think. After all, my brain needs fuel to come up with a solution to the great Sunday lunch dilemma.

'Hey Oreo,' I ask, 'do you know what is common between a bag of potato chips, a silicon microprocessor and a dwarf elephant?' This isn't the lead-in to a silly joke; these things are actually connected.

Consider the potato chip (if you're from the UK, you'll call them crisps) – invented, as legend has it, by a chef named George Crum in 1853 when an irritated customer complained his fried potatoes were too thick. Crum, in an act of culinary spite, sliced the potatoes paper-thin, creating what would become one of the world's most popular snacks. But the potato chip isn't just a tasty accident – it's an elegant solution to a fundamental problem: How to make food more efficient to store and transport.

By removing much of the water and creating an incredibly thin slice of potato, Crum had unknowingly created an early example of resource optimization. Through this transformation, a single potato could cover the surface area of an entire dinner plate. The chip's efficiency doesn't stop at space saving – its increased surface area allows it to cook faster with less energy, and its dehydrated state gives it a far longer shelf life than its unprocessed ancestor. The deceptively thin frame also fools us into thinking we're eating less.

As I munch on the chips, inspiration strikes. I shall do the opposite of what Crum did. I shall add more water and make my potatoes into a stew-like sabji – a *rasa*. It is a sabji traditionally made with the simplest ingredients – just potatoes, tomatoes, some spices and a lot of water – transformed into a hearty full meal.

This principle of efficiency through transformation mirrors what happened to elephants on Mediterranean islands millions of years ago. It's also found in another type of chip – the one in your laptop, and your phone – the silicon chip. Here's how.

Just as those island elephants underwent a fascinating transformation, becoming more compact while retaining their fundamental 'elephantness', the evolution of silicon chips too tells a tale of elegant miniaturization. Their precursors, the early integrated circuits of the 1960s, were like those first ungainly dwarf elephants taking tentative steps toward their new form: bulky, power-hungry and far from refined.

Through decades of relentless innovation, chip makers transformed from clumsy sculptors into atomic choreographers, directing a ballet in which the dancers are smaller than the eye can see. They mastered photolithography – which sounds like a fancy word for 'using light to draw stuff' because, well, that's exactly what it is. Think of it as the world's most expensive overhead projector, except instead of being used to project movies on a wall, it's used to beam patterns onto silicon wafers with the precision of a surgeon who's had way too much coffee.

The progress has been staggering. In 1971, semiconductors were being made at the 10 micrometre scale, which is about the width of a spider's silk thread. Today, they're down to 3 nanometres, roughly the length by which your fingernails grow in three seconds. It's like going from drawing with those chunky kindergarten markers to painting with individual atoms while wearing boxing gloves. In space. Blindfolded.

But this shrinking act wasn't just about making things smaller – that would be like breeding miniature elephants just because they're cute. Each new chip generation had to 'learn' to do more with less, like a rich minimalist who somehow fits an entire house's worth of stuff into a studio apartment. Engineers started stacking transistors in three dimensions, playing atomic Tetris with components so small they'd make a dust mite feel like Godzilla.

They also had to get clever about controlling those pesky electrons, which have the annoying habit of leaking faster than gossip at a family reunion. New materials and logic gate designs were developed to keep these subatomic troublemakers in line – imagine building a fence for particles that can literally teleport through walls (yes, electrons can actually do that, because quantum physics apparently runs on cartoon logic). The result? Chips that run cooler than a polar bear's toenails while performing calculations that would make a 1970s supercomputer go into an existential crisis.

And just as those island elephants unmistakably remained elephants despite their diminutive size, modern chips too maintain the fundamental logic of their ancestors – they're just extraordinarily more efficient at it. A modern smartphone chip that fits on your fingernail can outperform a room-sized computer from the 1970s while using a fraction of the energy, much like a dwarf elephant that has shrink-wrapped and preserved all its elephant capabilities.

I add a small amount of oil, cumin, crushed tomatoes, turmeric and salt to a pan, followed by the boiled potatoes, crushing them instead of cutting them in even cubes. Having improvised a similar curry like this once when I couldn't find a knife, I found that the flavours were richer when the potatoes were of uneven sizes.

Sometimes, happy accidents like this are a reminder how imperfections can actually lead to breakthroughs. Nowhere is this more evident than in the story of semiconductors. Imperfections paved the way for our modern-day silicon chips that can work on a nanoscale level. This control at the atomic and molecular scale was not just a technical achievement but a complete reimagining of what materials could do.

For much of history, materials were divided into two categories: conductors, which allowed electricity to flow easily, and insulators, which resisted it. Then semiconductors emerged. Crystalline structures like silicon and germanium occupy a remarkable middle ground, whose electrons can cross a small energy gap if provided enough energy from heat or light.

In their pure state, however, semiconductors were limited in practical use. At room temperature, they conduct poorly, making them inadequate for building reliable electronic devices. The true breakthrough did not come from the materials themselves but from the ability to precisely alter them.

In the late 1940s, researchers at Bell Laboratories discovered that by introducing small amounts of specific impurities into a semiconductor's crystal lattice, they could dramatically change its electrical properties. This process, known as doping, involved adding elements such as phosphorus, which contributes extra electrons, or boron, which creates vacancies called holes. Through this careful modification, scientists could control the flow of charge carriers with remarkable precision.

Doping made possible the creation of p-n junctions, where regions of positive and negative charge meet. These junctions became the fundamental structures underlying diodes and transistors. They allowed for the control, amplification and switching of electrical signals, forming the essential building blocks of modern electronics. Without doping, the transistor would not have been possible, and without the transistor, there would be no integrated circuits, no computers, no digital communication.

Here's another story closer to home: In the late 1950s, a modest laboratory at the Tata Institute of Fundamental Research (TIFR) in what was then called Bombay became the site of a remarkable story of scientific ingenuity. Physicist Homi Bhabha pioneered the building of India's first indigenous computer, the TIFRAC (TIFR Automatic Calculator).

R. Narasimhan was a key scientist who had just returned from the University of Illinois, where he witnessed the power of early computers like ILLIAC I. Inspired by their potential, he brought back more than knowledge; he also brought back a dream to build a computer entirely from scratch in India. What made this endeavour extraordinary wasn't just its ambition, but the sheer number of constraints under which it was achieved.

In the aftermath of Independence, India faced strict import restrictions and limited access to advanced technology. The scientists at TIFR were up against questions they'd never had to answer before. For instance: What makes the perfect cleanroom filter? Today, the answer involves sophisticated HEPA filters and complex air systems. But in 1960s' India, the solution was far less conventional. Anecdotally, it was something they rigged with locally found materials. While there's no concrete evidence confirming this, the story lingers as an enduring example of resourcefulness – whether fully true or not, it symbolizes the spirit of innovation.

What we do know beyond doubt is that these scientists had to become chemists, physicists and engineers all at once. Imagine being one of them, working with semiconductors whose dopant concentrations were minuscule – sometimes just one extra atom among millions. If you were building transistors, measuring those concentrations was not just important but also critical. A slight deviation could drastically alter the behaviour of the material.

However, accurately measuring dopant levels posed a major

challenge. These impurities were introduced at such tiny scales that even small variations could lead to unpredictable results. In the absence of advanced tools, the scientists at TIFR improvised and adapted methods like using electrical resistance, four-point probe measurements and Hall Effect[2] measurements to analyse dopant concentrations in semiconductors. Each step required painstaking improvisation and meticulous record-keeping. The TIFRAC team proved that technological breakthroughs aren't just about expensive equipment or imported components, they're also about human ingenuity and the willingness to try to do the impossible.

Indian science has long been defined by a kind of sustainability – not the Instagram-friendly kind, but the gritty, 'we have no other choice' kind. Long before sustainability became a buzzword, Indian scientists were figuring out how to stretch limited resources to achieve world-class results.

A version of this style of improvising based on scarce materials is called jugaad and it features in papers written by professors at Harvard Business School.

To be clear, though, when it comes to doing science under constraints, jugaad involves achieving extreme levels of precision and being able to repeatedly do marvellous things while relying on very few ingredients. This is miles (or kilometres) apart from sticking a piece of paper under a table to stop it from wobbling.

---

[2] A Hall Effect is a process used to develop an electrical field within a material that is oriented perpendicular to the direction in which an electrical current is flowing. It's useful in understanding and measuring the strength of magnetic and electrical fields.

Jugaad is how the TIFRAC team came up with bio-compatible electronic components and recyclable semiconductor materials. Necessity isn't just the mother of invention, it's also the strict aunt who insists on squeezing every last drop out of the toothpaste tube.

This resourcefulness isn't unique to modern laboratories. It's woven into the very fabric of civilizations that faced scarcity. Towards the east of India, the Hooghly River flows quietly through Bengal, its waters carrying centuries of stories, whispered in poems and sung by boatmen. Along its banks lies something seemingly mundane: grains of sand. If you lift it in your hand, you may think you're holding nothing. Nothing of note, anyway. But here's the twist – this sand is rich in silica, the foundation of some of the purest silicon in the world. Imagine this: The brains of your smartphone, the semiconductors powering entire industries, might have started life as a humble grain of sand dreaming big. It's yet another example of how sometimes what we see as waste may contain treasures that get unearthed with a bit of imagination, and sometimes a motivation due to scarcity.

The Hooghly has witnessed more than just technological ambition; it's seen resilience etched into the daily lives of its people. Take posto, for example. Aloo posto is one of my favourite dishes, and I hogged it enthusiastically any time my friend's mom made it. The story of how it came to be is a dark one, but also one of hope.

Back in colonial Bengal, the British, in their infinite pursuit of profits, forced farmers to grow opium instead of rice and other staples. This led to devastating food shortages. But Bengali women did what women have done for centuries.

They fed their families. Armed with little more than ingenuity and a grinding stone, they looked at the leftovers of the opium trade – poppy seeds – and thought, 'We can work with this.' What was once considered waste became a lifeline. These Bengali women turned tiny seeds into culinary gold: aloo posto and posto bora (a vada made of opium seeds), simple, nourishing dishes that have since become Bengali staples. Scarcity, it seems, can lead to something incredibly tasty when paired with creativity.

Further south, another story of scarcity unfolded in Bangalore during World War II. With rice supplies dwindling, the famed restaurant Mavalli Tiffin Rooms (MTR) faced a crisis: How do you make idlis without rice? Yagnanarayana Maiya, the founder, didn't panic. Instead, he reached for semolina, a grain sitting quietly in the pantry, and invented rava idli. It wasn't just a replacement – it was a reinvention, a dish so fluffy and delicious that it became an icon in its own right. Necessity didn't just mother invention here; it practically hosted a cooking show.

And then there's Korea, under Japanese colonial rule in the early twentieth century. Rice – the heart of every Korean meal – was being shipped out to Japan, leaving the locals with barely enough for themselves. But Koreans are nothing if not resourceful. They began mixing rice with barley (*boribap*) or millet to stretch what little they had, creating hearty meals that kept hunger at bay. Later, dishes like *budae jjigae* (army stew) emerged from post-war scarcity, blending scraps like Spam, sausages and kimchi into a pot of soul-warming comfort. It's fusion cuisine, born not of luxury but survival.

These stories of scarcity remind me of my childhood

where leftovers weren't just reheated – they were reimagined. Yesterday's idlis, cold and stiff, weren't tossed out; they were transformed. Sliced into neat little pieces, fried with mustard seeds, curry leaves and chilies, they became a breakfast that rivalled the original. Often, even when I didn't want to eat the idlis themselves, I'd look forward to the fried ones for breakfast. When I cried and insisted that I didn't want to eat, my father would take matters into his own hands. For him, stale rotis weren't a disappointment but an opportunity. Torn into bits and cooked in warm milk with a sprinkle of sugar, they turned into a dessert so satisfying that I ate them with glee thinking a whole new dessert had been cooked just for me. As a family that always prioritized reducing wastage, reimagined leftovers were proof that every meal, no matter how humble, had the potential to shine a second time.

As I continue on my cooking adventure, now whipping up some coffee, doubling down on my conviction to cook something from whatever I had on hand, I am reminded of a time that threw me into isolation – where scarcity was not food, but connection.

When the pandemic hit in 2020, my world turned upside down. One day, it was theatre rehearsals, research deadlines and the constant hum of human interaction. The next, I was alone in my first-ever rented apartment, staring at blank walls and grappling with the kind of silence that felt deafening. Moving from the lively chaos of a hostel to this solitary space was supposed to be a declaration of independence. Instead, it felt like I'd been marooned on an island, cut off not by oceans but by the absence of people, routine and certainty.

I knew, of course, that my struggles came with privilege.

The worst thing I faced was being alone, while so many others dealt with unimaginable losses – of loved ones, of jobs, of security itself. My loneliness didn't compare to the battles fought by those on the front lines, or those without the safety net I had. I had a roof over my head, and the ability to work remotely: These aren't things I took lightly. But even privilege doesn't shield you from the strange ways your mind can unravel when you're isolated from everything you once took for granted.

Then came Oreo. Or rather, Oreo came for me. When I went to look at him, I was still unsure if I was ready to take on the responsibility. I had a hard time looking after myself; was bringing a puppy into the mix fair? Oreo made the decision for me. This tiny black-and-white bundle of mischief marched up to me, locked eyes and essentially said, 'You're mine now.' If you've ever seen his giant eyes, you'd know I had no choice. And just like that, I wasn't alone anymore.

I had no idea what I was in for. Initially, Oreo was scared, and I was scared. He was always looking at me for comfort, barely making a noise. But then he got comfortable, and boy oh boy did he get comfortable! Suddenly, Oreo wasn't a quiet, calming presence; he was chaos in fur, a four-legged tornado with boundless energy and an unshakable determination to be in charge. He barked at invisible ghosts, chewed through furniture and took great delight in turning my apartment into his personal obstacle course, finding inventive ways to get into the kitchen. For someone who was barely figuring out how to live alone, adding a mischievous dictator to the mix felt like inviting trouble. And yet, he was exactly what I needed.

Oreo didn't just disrupt my days, he rewired them. One

of his first acts of genius was teaching me how to eat. Left to my own devices, I often forgot meals, too caught up in work or just the haze of pandemic-induced monotony. But Oreo wasn't having it. If I stayed at my desk too long, he'd start with a gentle nudge, then paw at my chair and finally resort to nibbling at my feet like a miniature drill sergeant. Once he got me up, he'd herd me toward the kitchen with all the focus of a sheepdog managing an unruly flock.

At first, I thought he just wanted to be fed, but even if I gave him food, he'd still sit impatiently, staring at me with his big puppy eyes, his ears stacked on top of his head and his head tilting from side to side at the annoyance of me just not getting it. And there he'd keep sitting, staring at his food bowl and then at me, as if to say, 'Your turn, human.' Only when I grabbed something to eat myself – be it leftovers, a snack or a hastily assembled sandwich – would he finally jump into his meal, sighing at the stupidity of the human who just doesn't learn, probably thinking, 'This one's gonna be hard to train!'

It was as if he'd taken it upon himself to make sure I didn't starve, if only to continue having someone to give him his next meal. It became a ritual, one that grounded me when everything else felt unmoored. In hindsight, it was Oreo who taught me the importance of routine, of finding anchors in the storm.

But the herding didn't stop with meals. If I spent too long staring at a screen, lost in work or doomscrolling, Oreo would intervene. He'd circle me, pawing and nudging until I stepped away, usually toward the couch, where he'd plop down in my lap, sigh and turn his head upside down like an owl and look triumphantly, as if to say, 'Good. Now breathe. Also, move a bit so I'm more comfortable.'

Oreo's antics were more than just amusing; they were transformative, a lifeline in what felt like being surrounded by the endless sea. Isolation had stripped away everything I thought defined me – work, performances, the endless buzz of people – and forced me to confront myself. But it was Oreo, the tiny pup who claimed me, who became my guide through the chaos. He turned an empty apartment into a home, isolation into connection and ordinary days into something extraordinary. In his own chaotic, herding way, Oreo taught me that sometimes, the best lessons come from a dog-shaped drill sergeant with a wagging tail and the determination to get you out of your chair, even if it means nibbling at your toes. For my part, I count it as one of my life's greatest achievements that I created a safe space for the scared little boy to turn into the demanding dictator he is now.

In a way, the story of the women trying to feed their families in difficult times mirrors the story of the scientists trying to build in scarcity, which mirrors the story of miniature elephants, potato chips and maybe even me. It's a story of making do with what you have, of finding brilliance in leftovers and of turning scarcity into fuel to innovate – whether we call it an innovation or not. Whether it's a grain of sand, a handful of poppy seeds, a scoop of semolina or elephants, these stories aren't just about survival but about transformation. It's ingenuity, a spark that turns what we have into what we need.

Here's the twist, though, because of course there is! Islands don't always make things smaller – and this is where the story gets wild. While some creatures downsize, others take the opposite route and go full 'go big or go home'.

Enter stage left: the Komodo dragon, the world's largest lizard, stomping around Indonesian islands like it owns the place. With no big cats or wolves to compete with, these lizards decided, 'Fine, I'll do it myself,' and claimed the title of top predator.

Then there was the dodo, a giant, flightless pigeon from Mauritius that looked at the sky and said, 'Nah, I'm good.' With no predators to run from, being grounded wasn't a problem, and its size helped it access food other birds couldn't. Meanwhile, the giant tortoises of the Galápagos and Aldabra took things even further, lumbering around their islands as if to say, 'Try messing with me, I dare you.' Spoiler: nothing did.

In fact, here's something to chew on. Remember how our beloved island elephants followed the familiar script of miniaturization? In the tight confines of small islands, where food was limited and survival demanded efficiency, shrinking made sense. Evolution, it seemed, was predictable – rule-based, even. However, the one thing that you must know is that if evolution has a rule, it is this: there is no rule without an exception.

Enter the elephants from the island of Sri Lanka. There, the elephants (*Elephas maximus maximus*) did not shrink. Despite being isolated from mainland Asia for tens of thousands of years, they remained resolutely massive. Adult males still tower up to 3.5 m at the shoulder and weigh over 5,000 kg, often outmatching their mainland ancestors in size and stature.

So, what happened? Several factors tipped the balance in their favour. Unlike the rocky outposts of the Mediterranean,

Sri Lanka is a relatively large island, rich in rainforests, grasslands and wetlands that provide abundant food year-round. Resources never became scarce enough to reward a smaller, more frugal body. Just as importantly, the island's ecosystems lacked major predators that would otherwise have pushed elephants to evolve faster reproductive cycles and smaller, more manoeuvrable bodies.

Genetic studies suggest another hidden advantage: Sri Lankan elephants maintained substantial genetic diversity over generations, shielding them from the kind of evolutionary bottlenecks that can accelerate size changes in small, isolated populations. This pattern of gigantism flips the usual story of island evolution on its head. While some species shrink to survive on limited resources, others take advantage of predator-free environments and stretch to their full potential. These oversized creatures fill roles that would never be available on the mainland, where predators like lions or tigers would keep them in check. It's as if islands give evolution a blank slate and say, 'Go ahead, get weird!'

Nowhere is this clearer than around Wallace's Line, one of the most dramatic natural divides in the world. Named after Alfred Russel Wallace, this invisible boundary slices through Indonesia, splitting Bali and Lombok, Borneo and Sulawesi. On the Asian side of the line, you'll find classic megafauna like tigers, elephants and rhinos – what you'd expect from a continent. But cross the line, and you're suddenly in a world of marsupials, cockatoos and giant rats that look like they belong in a fantasy novel.

Why the stark difference? During the ice ages, when sea levels dropped, the islands on the Asian side connected to the mainland, forming a massive landmass called Sundaland. Animals could roam freely, and ecosystems stayed more or less consistent. But on the other side of Wallace's Line, deep ocean trenches stood their ground, isolating the islands no matter how low the seas fell. This separation lasted for millions of years, giving evolution time to experiment.

And experiment it did. With no lions, tigers or elephants to share space with, the islands east of Wallace's Line became evolutionary playgrounds. Birds stepped up to fill roles usually reserved for mammals, like the cassowary, which looks like it should be guarding the gates of a lost kingdom. Rodents grew to sizes that would make you rethink the phrase 'tiny as a mouse'. Marsupials, which had largely disappeared from Asia, flourished in these isolated ecosystems, becoming the stars of their own evolutionary dramas.

This stark division underscores a broader truth: Isolation doesn't just shape ecosystems, it also reshapes life itself. Darwin's famous finches on the Galápagos Islands are a perfect example of this principle. From a single ancestral species, these birds radiated into multiple species, each with distinct beak shapes tailored to different diets. It's evolution in action, a story repeated in Hawaiian honeycreepers, Madagascar's lemurs and the fruit flies of the Seychelles. Islands force species to get creative, and nature, ever resourceful, rises to the challenge.

As I get the flour out to knead the dough, I find half a beetroot. I'm in the mood to experiment, so why not add this to my dough?

This story isn't confined to biology. The same pattern – where isolation drives innovation – emerges in technology, too. Take supercomputers, for example. These aren't just oversized PCs; they're the tech equivalent of Komodo dragons, growing massive in their isolation. Programmes like the IBM Summit, housed in a climate-controlled fortress at the Oak Ridge National Laboratory in the USA, were built to solve problems too big for ordinary machines. Capable of performing over 200 quadrillion calculations per second, Summit was a computational juggernaut designed to tackle global challenges like climate modelling and genomics. While we used our laptops to Google 'Why is my plant dying?'[3], Summit was out there crunching data to save the planet.

Then there are offshore oil rigs, floating islands of technology out in the deep blue. These rigs are like the giant Komodo dragons of the sea – massive, self-sufficient and a little intimidating.

Saturation diving rigs weren't built because someone thought it might be fun to live in a giant pressurized can at the bottom of the ocean. They were built because the sea said no, and humans, being humans, decided to negotiate. In the mid-twentieth century, as oil and gas companies pushed their drilling platforms into deeper waters, sometimes over 300 metres deep, where pressures can exceed 30 times what we experience at the surface, traditional diving methods began to fail. Divers could only dip into the depths for fleeting minutes before the crushing weight of the ocean and the treacherous physics of decompression sickness demanded they retreat. It was inefficient. It was risky. It was not good for business.

---

[3] The answer is always 'because you didn't water it ... or watered it too much … or looked at it wrong'.

Of course, the story of offshore drilling and its enormous, multifaceted consequences for the planet is a conversation for another day. But within that complicated history, this much is clear: The demands of the deep forced us to invent new ways of living at the edge of survival.

Thus, saturation diving was born. Entire floating cities had to be designed: Ships over 100 m long, carrying pressurized living chambers that could house a dozen divers at a time, transfer capsules built to withstand external pressures of 300 psi, life support systems that could continuously mix and monitor helium–oxygen gas blends, hyperbaric medical bays, emergency recompression units and redundant backups for every vital system. Inside these vessels, divers lived at pressures mimicking the seafloor for up to 30 days without ever surfacing. Staying alive underwater became a full-time engineering project, a careful negotiation of every breath, every heartbeat.

Isolation forced engineers to make these large underwater rigs ridiculously robust because they have to be able to handle everything from angry waves to relentless weather to extracting oil from way beneath the ocean floor. These rigs don't just extract resources; they're proof that when the going gets tough, humans build bigger and stronger.

Space exploration might be yet another example. We build large, sprawling, contained structures allowing us to survive in harsh environments not meant for us to live in. Enter the International Space Station (ISS), a tech fortress orbiting about 400 km above Earth. Unlike satellites, which are small and neat, the ISS is a sprawling modular giant. Stretching 109 m from end to end, roughly the size of a football field, it is huge because, well, it has to keep humans alive in the middle of space. This technological juggernaut wouldn't exist if isolation didn't demand something grand and ambitious.

And we can't forget data centres. If the ISS is a space-based giant, data centres are its earthbound cousins. These digital islands house thousands of servers across millions of square feet, storing everything from your photos of latte art to entire Netflix series. Here's a secret: There's no such thing as a 'cloud'. When your data is stored 'on the cloud', it's really stored on a data drive in a remote location.

Facilities like the Lakeside Technology Center, covering around 1.1 million sq. ft (about 25 acres), in Chicago are so massive that they make your phone's memory look like a sticky note. Why are they so big? Because the scale of these systems demands it. They're designed to handle insane amounts of data. The situation is reversed here – data centres need to be in isolated areas because of how big they are and also because they need to be in large, secure facilities. It's funny, though, how things that start as practical solutions to a problem can evolve into something so grand they almost forget their humble origins. As we enter into more and more cloud computing and into the age of AI, data centres continue to grow bigger and flashier because they have to.

But this pattern isn't exclusive to technology; it's been simmering in kitchens for centuries. Case in point: biryani.

Ooh. Biryani! Should I order some? No. I will make food with whatever I have. But coming back to biryani …

Biryani began as the culinary equivalent of a data centre – built for function. Soldiers needed hearty, easy-to-make meals, so someone thought, 'Throw some rice, meat and spices into one pot, and let's call it a day.' Efficient, nourishing and practical. But then things escalated. Today, biryani is served with saffron, caramelized onions, boiled eggs, nuts, potatoes (in Bengal) and occasionally an air of superiority. What started as a one-pot meal for survival is now a dish so extravagant it feels like it should come with a red carpet.

And it's not just biryani. Over in Spain, paella followed a similar path. Farmers tossed rice, beans and whatever meat or fish was lying around into a pan over an open fire. Simple, right? But then someone decided, 'What if we add saffron? And prawns? And artfully arrange everything so it looks like a painting?' Now, paella is a global icon, served at parties where the only leftovers are bragging rights.

Or take Louisiana's gumbo. It started as a practical way to incorporate scraps of seafood and meat with okra or roux into a meal that could feed a family. Fast-forward a few decades, and gumbo has become the star of festivals and cook-offs, with chefs battling to see who can make the most elaborate, flavour-packed version. Scarcity met creativity, and the result was pure delicious drama.

Closer to home, there's nihari, that slow-cooked stew that began as breakfast for labourers. Long hours of simmering turned cheap cuts of meat into something tender and comforting. But someone, somewhere, looked at that pot of stew and thought, 'This needs marrow. And ginger. And garnish worthy of a Mughal emperor.' Today, nihari is served with such a flourish that it feels like you're being handed an heirloom recipe every time it's placed in front of you.

And here's the thing: This isn't just about food – it's about what scarcity does to the human spirit. Necessity doesn't just breed invention; it also breeds imagination. It turns 'What can we do with this?' into 'How far can we take this?' Whether it's a data centre storing memes or a biryani layered with saffron, the narrative is the same. We don't just survive – we thrive, and we do it with flair.

Inspired, I decide to add some homemade garam masala mix to my curry. A simple addition, but it transforms the dish into something divine, a sum definitely greater than its parts. Nature plays this game too. Isolation doesn't always lead to miniaturization. Look at the Komodo dragon, a lizard the size of a horse, ruling its island kingdom. Or take supercomputers, which started as room-sized machines and have grown into climate-controlled monsters capable of eating math problems for breakfast. Whether it's nature or technology, isolation has a way of creating giants – sometimes literally.

But going big comes with its own supersized problems. Komodo dragons need huge territories to hunt and can't exactly hop between islands for a quick snack. Supercomputers guzzle electricity in the same way that teenagers raid fridges, requiring entire power plants to keep them running. It's the same headache whether you're a massive lizard or a massive machine – the bigger you are, the more resources you need just to exist. And don't even get me started on cooling systems – both Komodo dragons and supercomputers spend a ridiculous amount of energy just trying not to overheat. It's like nature and technology independently discovered that being huge means spending half your time dealing with infrastructure problems.

Getting smaller comes with its own set of size problems. When elephants decided to try island life and go compact (not exactly their choice, but whatever), they had to deal with more than just fitting into smaller jeans. Their entire body had to figure out new ways to work – like someone moving from a mansion to a studio apartment and realizing that no, the grand piano won't fit through the door.

Similarly, when chip makers try to squeeze more transistors into tinier spaces, they run into the problem of quantum physics – which is basically nature's way of saying, 'You're not the boss of me.' Electrons start getting as unpredictable as cats in a laser light show; they're supposed to stay in their lanes, but suddenly start popping up everywhere they shouldn't be. (That's quantum tunnelling for you – as bizarre as it sounds, it's very real and very annoying for chip designers.)

Here's where Moore's Law comes in – the famous observation that the number of transistors on a chip doubles about every two years while costs get halved. This 'law' is not really a law. It's an empirical observation made in the 1960s by Gordon Moore, who was one of the founders of Intel. It has held true for decades, though nowadays it's starting to show signs of strain. Turns out even exponential growth has its limits.

Strangely, as island elephants developed thicker legs to support their new, more compact bodies, their brains actually got bigger relative to their size. It's like they were following Moore's Law before it was cool, getting smaller while processing more information. Meanwhile, modern chips keep getting more powerful while shrinking, which is great until you hit the atomic scale and physics starts throwing tantrums.

Still, there's poetry in these constraints: Just look at how the same genetic machinery[4] can create both miniature and massive forms, echoing how human ingenuity under scarcity leads to both impossibly small chips and grandiose technological solutions. Here's something I've realised: The universe is made of patterns. Once you spot them, they have the nasty habit of showing up at every corner. Whether it's food, technology or evolution, the same stories keep getting told with different characters. The tale of 'doing more with less' appears everywhere, from saffron-laced biryani to Komodo dragons, from supercomputers to gumbo. The details change, but the rhythm of the story stays remarkably consistent.

This isn't just a quirk of biology or engineering; it also serves as a crystal ball that shows us our future. Human activity has carved up the planet into fragmented habitats surrounded by urban 'seas'. These spaces function like islands, forcing species to adapt under intense pressure. Speaking of which, elephants continue to adapt in unusual ways. Asian elephants demand a sugarcane tax by stopping trucks in Indonesia, and demand corn in Sri Lanka. For giants that need more than 300 kg of food per day, stuck in a situation where we artificially reduce their movement and access to food, this is a clever way to do jugaad, but I'm not sure it's enough.

And here's the beautiful irony: The lessons of scarcity and isolation come full circle. The same principles that shrank elephants to the size of dogs also made Komodo dragons massive, biryani extravagant and supercomputers colossal.

---

[4] The idea of genetic machinery as mentioned here is gross oversimplification. It's quite complicated really, but honestly, if you break it down, most machinery is complicated.

Scarcity forces ingenuity, whether it's the invention of rava idli during a rice shortage, or my dad turning yesterday's leftovers into something so delicious it feels like a planned masterpiece. These forces of adaptation, whether they lead to miniaturization or grandeur, show us how limitations can spark creativity and transform the ordinary into something extraordinary.

So, whether it's tiny elephants or giant lizards, eco-friendly ingenuity or the magic sand of the Hooghly, a one-pot meal turned into a saffron-laced masterpiece or a cold idli reborn as a crispy, spiced delight, resource scarcity and its impact on the world have been anything but small. Sometimes, the smartest solutions emerge from squeezing the most out of the scantiest resources – or dreaming big in the face of limitations – all with a healthy dose of stubborn creativity.

I, for now, will go eat this meal that I've cooked up with limited ingredients, following in the footsteps of so many ingenious women before me. My roti is pink from the beetroot, and my curry smells like nostalgia and self-reliance. Granted, my scarcity is entirely self-imposed (because I've chosen not to buy more ingredients), but it's still turned out to be surprisingly delicious!

6

Breaking Good or the Chaos Monkey Paradigm

After lunch, it is time for cleanup – one of the most painful realities of living alone. I turn to look at Oreo to ask if he's going to help with the cleanup but, of course, he just looks back saying, 'I can add to the mess if you want.' I quickly start cleaning up, because any cleaning process is controlled destruction, and Oreo has the destruction part down pat, but not so much the controlled part.

As I start moving vessels into the kitchen sink, I encounter a stubborn stain that refuses to budge. But I bring out my secret weapon – the unassuming simple dish soap. With just a little bit of soap and water, the stain starts breaking down, almost like magic. In fact, at the microscopic level, soap is doing something rather fascinating, which to me is similar to magic.

Soap, you see, is one of the most efficient methods of controlled destruction, rivalled only by my mother's biting words when I oversleep. Imagine you're watching a demolition team at work, except this team is so small you'd need a microscope the size of a house to see it. That's basically what soap molecules are doing right now on my kitchen counter. They're like highly specialized workers – one end of the molecule, known as hydrophilic head, loves water, while the other end, the hydrophobic tail, is more interested in hanging out with grease and dirt. This split personality makes its incredibly good at its job.

Water alone is not sufficient to dissolve dust and grime because water molecules stick with each other better than they stick with molecules that have fats (and what we describe as grime usually contains fat). It is the same with bacteria, which have a fat layer on the outside protecting the inside from water or any other similar solvent.

Here's how soap does its magic. When these molecular demolition experts meet bacteria, they don't just politely ask them to leave. Oh no! They go full action movie star on them, literally ripping them apart. The water-loving parts grab one side, the grease-loving parts grab the other, and *poof* – bacterial cell membranes are torn apart faster than I devour a hot dosa after a long hike (or, actually, the way I devour dosas any time).

Even viruses, which are immune to antibacterial agents (which is why there's no point in taking antibiotics when you have a viral infection), do not stand a chance in front of the mighty soap. When you wash your hands with soap, you're essentially unleashing millions of tiny demolition experts on these viral villains. The soap molecules essentially grab onto the virus's fatty outer layer and literally pull it apart. It's like someone trying to wear a jacket while two people pull it in opposite directions. That jacket's not going to survive, and neither does the virus. The only thing it needs is a little bit of time. This is exactly why doctors kept telling us during the pandemic to wash our hands for more than 20 seconds at a time.

However, just like most things in life, too much of anything can be overkill.

When we overuse soap, we're not just killing harmful bacteria but also disrupting a complex ecosystem that's living on our skin. Our skin hosts millions of beneficial bacteria that actually help keep us healthy by preventing harmful microbes from gaining a foothold on our skin.

(I do not like the nomenclature of 'good and bad' bacteria, but it will serve as shorthand. The same bacteria that are beneficial in our gut can turn dangerous if they get into the bloodstream, so there's no 'good' bacteria, just bacteria that are beneficial to us in the context they are in. 'Harmful' bacteria are also harmful only in context.)

These good bacteria are an essential part of our body's defence system. They produce substances that keep harmful bacteria in check, compete for resources with potential pathogens and help maintain your skin's slightly acidic environment that many harmful microbes hate.

But soap is an indiscriminate destroyer. Every time we use it, we're not just removing dirt and harmful microbes; we're also impacting this beneficial bacterial community. Even worse are the abominations known as the antibacterial soaps. While regular soap physically breaks apart microbes (that cool demolition process we talked about earlier), antibacterial soaps contain additional chemicals specifically designed to kill bacteria. The most common one used to be triclosan, until we realized it was about as precise as using a grenade to open a jar.

These antibacterial chemicals don't just kill harmful bacteria but can create something way worse: resistant bacteria. It's like when you don't finish a full course of antibiotics and the surviving bacteria come back stronger. The bacteria that survive the antibacterial assault can develop resistance, turning into the microbial equivalent of supervillains.

The kicker? Studies show that antibacterial soaps aren't any better at preventing illness than regular soap. Regular soap's physical demolition method is already incredibly effective. Adding antibacterial agents is like hiring a wrecking ball operator who also brings their own explosives – unnecessary and potentially counterproductive, unless you're shooting a classic movie like *Fantastic Voyage*.

This is why scientists and healthcare workers often stick to regular soap. They understand that sometimes the simplest solution is the most elegant – you don't need to carpet bomb your skin's ecosystem when a precise strike will do the job.

This is not to say that antibacterial agents do not have a place. It's all about balance and using the right tool for the right job. Each has its place, its purpose, and trying to use the nuclear option for every situation is not just overkill but can actually make things worse.

Hospitals have figured out this categorization perfectly. In the lobby? Regular soap. In patient wards? Slightly stronger stuff. Before surgery? That's when they bring out the serious antimicrobial agents because, in that context, you really do want to temporarily create a near-sterile environment.

But – and here's the crucial part – bringing that industrial-strength cleaning power into your home bathroom is like using a pressure washer to water your houseplants. You'll definitely kill any pests, but you'll probably kill the plant too.

This is why microbiologists often roll their eyes at 'antibacterial' everything. In most daily situations, regular soap achieves the perfect balance – it's strong enough to break apart the microbes you don't want while being gentle enough not to completely devastate your skin's microbial ecosystem.

Even the often-retold story of how soap came about is itself a story of controlled destruction. According to Roman legend, soap was discovered on Mount Sapo, where animal sacrifices were regularly performed and holy fires were lit. Rain would wash the animal fat and wood ashes down the hill, and women washing clothes in the Tiber River below noticed something strange – their clothes got cleaner when they washed them in certain spots.

What was happening was a perfect chemical accident. The fat from the sacrifices mixed with alkaline wood ashes, brought to the river by rainwater, helped trigger the reaction we now know as 'saponification'. The women at the river had stumbled upon the world's first soap factory – courtesy of rain, religious ceremonies and pure chance.

When you mix fat (like the animal fat in this case) with a strong base (wood ashes in the example above), something remarkable happens. The fat molecules get split apart in a process that's basically molecular divorce. The base grabs the fatty acid chains and converts them into something new – soap molecules and glycerine (which is the reason why handmade soaps are often more moisturising than commercial soap).

This reaction is a bit like choreographed demolition – precise, controlled and transformative. The base acts like a wrecking ball, breaking apart the fat molecules at exactly the right points. Instead of just creating rubble, though, it builds something new and useful. The resulting soap molecules have that split personality we talked about – one end that loves water and one end that loves grease.

Think of it like taking apart a car and using its parts to build a submarine. You're not just destroying something, you're creating a new thing with completely different capabilities. The end product can travel in environments (like water) that the original fat molecules couldn't navigate effectively.

The Roman origin story may well be apocryphal and just a story, embellished in many retellings where historical accuracy was not a concern. But here's the thing – while the Romans left us this great origin story, they weren't actually the first to figure out soap.

The earliest known soap recipe was written on a Babylonian clay tablet around 2800 BCE. It called for water, an alkali (potash) and cassia oil. The Babylonians probably used it to clean wool and cotton for textile production rather than for personal hygiene.

The ancient Egyptians were also players of the soap game. Medical documents from around 1550 BCE describe combining animal and vegetable oils with alkaline salts to create a soap-like substance. They used it for treating skin diseases and for washing.

Of course, this also takes me back to Sundays at home when my mother used to boil reetha, also known as soapnut, to wash my hair after intensely oiling it. As I remember sitting cross-legged on my bedroom floor, with my mother working warm coconut oil into my scalp, I think about controlled destruction. Each strand of hair I'm nurturing is dead, a protein filament that was destroyed the moment it emerged from my scalp. Yet in its death, it serves a vital purpose – protecting my very-much-alive scalp from the sun, keeping me warm and, in more ancient times, providing sensory awareness through movement. After the oil massage, I would use crushed reetha and shikakai to wash the hair, which was meant to make the hair much softer. This practice has been followed in India for the longest time, with these plant-based soap molecules being used to wash clothes, hair and everything else.

While traditional soap comes from that fat-meets-alkali reaction we talked about, soapnuts contain something called saponins – naturally occurring compounds that happen to act surprisingly similar to soap. When you put reetha in water and agitate it, these saponins create that familiar foamy lather. They're like nature's pre-made soap molecules, complete with that split personality that makes them good at grabbing both water and dirt.

The really clever bit? Plants evolved these compounds as a defence mechanism. Saponins are actually toxic to many insects and microorganisms, which is why they make such effective natural cleaners. It's like the plant developed its own built-in pest control that we humans later discovered could clean our clothes, much like chillies evolved capsaicin, the compound that gives them their spicy taste, as a defence mechanism.

But here's where they differ from traditional soap: Saponins are structurally different from soap molecules. While both have that water-loving and water-hating end, saponins are more complex molecules – often, but not always having more complex molecular structures and activity levels. They're often gentler on fabrics and skin, which is why they've been used in Ayurvedic medicine and traditional cleaning for thousands of years. One of the key differences is that saponins by their sheer makeup start biodegrading fairly quickly, unlike artificially prepared detergents which are shelf stable for longer. The same properties that make saponin-based cleansers gentler (usually but not always), make them unsuitable for long-term storage.

Saponins are not unique to soapnuts. Other plants got in on the action too. Sisal, yucca and soapwort all produce different types of saponins. Each plant evolved its own slightly different version, much like different cultures developing their own soap recipes. Even among plant-based soaps, different cultures used different plants. Native Americans used yucca root, while ancient Greeks preferred soapwort and the larger South Asia used soapnuts. Same principle, different execution.

Now that the counter looks clean-*ish*, I must follow Oreo, who is sheepishly looking at me, telepathically communicating that we need to go and sit. Of course, he could just go away and leave me to work alone in the kitchen, but some sheepdog ancestry (unverified) in him insists that the sheep must be watched, and since there are no sheep in my flat, I am the stand-in.

Of course, who am I to argue with the maalik of the house? So we go in, getting comfortable on the sofa, and I ponder that very First-World problem: 'What should I watch?'

As I toggle through different streaming services, I remember the weird but aptly named 'Chaos Monkey' tool. Just like soaps cause chaos at the molecular level with their dual personalities, leading to controlled destruction, the Chaos Monkey causes controlled destruction to make algorithms better.

In their now-famous 'Chaos Monkey' (a nickname that was also often used to describe me in childhood), Netflix pioneered this approach. It was named after the idea of a monkey randomly breaking things in a data centre. But this wasn't random destruction; it was carefully orchestrated chaos. The Chaos Monkey would randomly shut down servers during business hours when engineers were available to respond, not in the middle of the night when failures could cascade uncontrolled. This practice forced development teams to build resilience into their systems from the ground up.

The concept evolved into the 'Simian Army' – a suite of testing tools each named for a different type of simulated destruction. The Chaos Gorilla simulates the outage of an entire Amazon Web Services availability zone. The Latency Monkey introduces artificial delays in network communications, mimicking degraded network conditions. The Conformity Monkey finds instances that don't adhere to best practices and terminates them. The Doctor Monkey taps into health checks that are being run to detect unhealthy instances and removes them from service.

Amazon took this concept further with their 'GameDay' exercises – entire days devoted to deliberately breaking things in controlled ways. Teams would gather in war rooms, ready to respond to artificially induced catastrophes. Google also developed DiRT (Disaster Recovery Testing), in which they simulate everything from data centre fires to earthquakes.

In the world of software development, controlled

destruction has evolved into a sophisticated practice where we keep deliberately introducing problems into a complex system – not to break it, but to make it stronger. Engineers start by forming specific hypotheses: 'If data centre A fails, users should still be able to watch their shows without interruption.' They define what 'normal' looks like for their system in terms of key metrics like response times and error rates. Then they introduce controlled disruptions and carefully measure what happens. It's like gaming an immune system challenge for technology.

This methodical approach turns seemingly random destruction into a carefully controlled experiment. Every test has specific conditions for stopping if things go wrong, much like how you'd stop exercising if you felt sharp pain rather than normal muscle fatigue. The goal isn't just to break things – it's to understand exactly how systems fail and ensure they can recover gracefully.

But why is this even needed?

Imagine you're building a house in an area that gets hurricanes. Instead of just hoping your house will survive when a big storm hits, wouldn't it be smart to test it against strong winds beforehand? You might use powerful fans to see if the windows hold up, or spray water to check if the roof leaks. That's exactly what big tech companies do with their digital systems, but on a much larger scale.

Think of a company like Netflix. When I press play on my favourite show, that simple action sets off a complex chain of events across thousands of computers (we call these data centres) all working together. If any part of this chain breaks, the show might buffer or stop playing entirely. Now imagine these processes happening for the millions of people watching at the same time! This is where 'chaos engineering' comes in – it's like having a safety inspector who deliberately causes small problems to prevent big disasters.

So, let me ask again: Why do this?

Because in the real world, things break all the time. A construction crew might accidentally cut through Internet cables. A storm might knock out the power supply for an entire building full of computers. A cosmic ray (yes, really!) might cause a bit flip in computer memory. By deliberately causing small failures, engineers learn to build systems that can handle such problems gracefully. It's like how your body builds immunity by fighting off small infections – the system gets stronger through controlled exposure to problems.

The end result? When real problems happen – and they always do, Murphy's Law is basically life's operating system – these systems are much more likely to keep running smoothly. That's why you can keep binge-watching your favourite show even if a massive storm knocks out power in another part of the country. Or why your carefully curated online shopping cart doesn't mysteriously disappear mid-scroll, even if a server somewhere gets tired and decides to take a break.

While this approach relies on introducing bugs to break down systems intentionally, the story of how the first ever bug came about comes to mind. Back in 1947, Grace Hopper, a pioneer in computer science, was working on the Harvard Mark II computer when something strange happened. The computer kept crashing, and no one could figure out why. When they finally opened up the machine, they found a moth trapped between the relay contacts. Hopper taped the moth into her logbook with the note: 'First actual case of bug being found.'

Today, this approach – accepting that things will break and preparing for it – has revolutionized how we build reliable digital services. Instead of naively trying to prevent every possible problem (an impossible dream, like staying awake during a three-hour meeting the content of which could actually have been compressed into a five-minute email), engineers build systems that expect failure. These systems heal themselves, reroute around issues and keep working, kind of like how your body does when you sprain an ankle but still manage to limp dramatically toward the fridge for ice cream[1]. Sometimes, the best way to strengthen something is to break it – but on purpose, and with control.

Controlled destruction appears in many forms all around us. Take metallurgy, for example. This ancient art is a prime example of destruction leading to creation. Much like our bones are constantly breaking down and rebuilding (yes, we're all living construction sites!) metal undergoes its own transformation through carefully controlled destruction.

Thanks to Kurosawa, Musashi and Hollywood, Japanese sword making gets all the hype. But ancient Indian metallurgists were doing some mind-blowing stuff long before Samurai movies were cool.

In South India, craftsmen perfected the creation of wootz steel, a material so legendary it became the basis for many stories around the world. It is the steel that was used to make

---

[1] As you can possibly tell, my metaphors, much like me, tend to be quite food motivated.

the famed Damascus blades – the ones that were the basis of the fictional Valyrian steel that is strong enough to slay dragons in the *Game of Thrones* series.

While the exact way the blades were made is lost to history, we do have some information on the steel itself. The steel was made by skilled metallurgists who heated steel ingots to just below their melting point (about 1,200°C, so, roughly the temperature of an argument on Twitter) in sealed crucibles containing plant materials and glass. The cooling process was equally precise, resulting in crystalline structures that gave the steel its incredible strength and distinctive water-like patterns. These blades weren't just weapons – they were art. Destruction and reformation carried out at a microscopic level, all in perfect harmony.

And then there's the Iron Pillar of Delhi, a 1,600-year-old metallurgical marvel. This rust-resistant wonder owes its durability to a clever process called phosphorization. Instead of rusting away like my first bike, the iron formed a protective oxide layer, shielding the metal underneath. It's the metal version of a skincare routine – controlled damage leading to long-term protection. Only, unlike the layers of skincare we apply on ourselves, the Iron Pillar's durability comes from a chemical process.

But wait, there's more. In Rajasthan's Zawar mines, ancient metallurgists were extracting zinc centuries before anyone else. They had to master really precise temperature control because the process of extraction was complex and exact. The zinc had to be vaporized (requiring a temperature of around 1,000°C hot, but not *too* hot) and then carefully condensed, all without wrecking the apparatus.

Even the act of eating is controlled destruction. Our stomach acid is so strong it could dissolve metal (seriously, hydrochloric acid is no joke), but it's confined within a carefully managed system. Food gets broken down into tiny components, which are then reassembled into energy and life. We destroy to sustain ourselves – literally.

And let's not forget construction. Traditional Indian builders used slaked lime (calcium hydroxide) in their mortar, which they made by burning limestone. But they didn't stop there. They added jaggery, fenugreek and other organic materials, which fermented over time and strengthened the mortar.

In my hometown Allahabad (now Prayagraj), there are buildings in Khusro Bagh which have ancient walls that are standing today, a testament to the quality of materials used. Once, during a picnic there, my parents introduced me to the ancient processes used to make these walls, sparking a series of curiosity-driven questions neither of them were ready to cope with.

Modern cities follow this same principle, often building directly on the ruins of what came before. Rome, for example, is layer upon layer of construction and destruction. In that way, the city itself is rather like a lasagna one may want to demolish when in Rome. Urban renewal today faces the challenge of balancing growth with preserving historical value, but the principle remains the same: break to build.

Even on a microscopic level, your cells are in on the act. The process of autophagy – literally 'self-eating', because biology and biologists like to be dramatic – is one of the most beautiful examples of destruction with purpose. In this process, your cells break down and recycle their own components, keeping you alive and functioning by repurposing what's no longer useful.

Autophagy is especially important during periods of fasting or stress, when your body taps into its reserves. Instead of panicking about the lack of external resources, it calmly assesses its own inventory and decides, 'This old, busted mitochondrion? We don't need that. Let's break it down and use the parts to keep things running.' This process allows your body to survive lean times by efficiently recycling cellular waste into usable energy and materials.

The elegance of autophagy goes far beyond survival mode. It's also a vital part of cellular housekeeping, clearing out damaged proteins and organelles that could otherwise accumulate and lead to diseases. It's like having a high-tech janitor constantly on duty, sweeping up cellular debris before it can cause trouble. Without autophagy, your cells would become cluttered and chaotic, overwhelmed by the biological equivalent of hoarding.

The significance of this process was recognized in 2016, when Japanese researcher Yoshinori Ohsumi was awarded the Nobel Prize in Medicine for his pioneering work in uncovering the mechanisms of autophagy. His research revealed how cells identify what needs to be recycled, package it into specialized vesicles called autophagosomes and deliver it to the lysosome – the cell's recycling centre – where it's broken down into reusable components. It's a process as precise and coordinated as any assembly line, and it's happening in your body right now.

What's even more amazing is how autophagy ties into

larger themes of health and longevity. Research suggests that stimulating autophagy through practices like intermittent fasting or calorie restriction may improve metabolic health, reduce inflammation and even slow the ageing process[2]. It's a reminder that, sometimes, less really is more: by giving your body a break from constant food intake, you allow it to focus on internal maintenance – extending its warranty, so to speak.

Autophagy doesn't just keep us alive; it's also a testament to the beauty of destruction. By breaking down what's no longer useful, our cells make way for renewal and resilience. It's a quiet, ongoing act of creation through destruction, happening trillions of times a day in every corner of your body. If that's not proof that biology has its own poetry, I don't know what is.

The same principle applies to art restoration, where destruction becomes a tool for uncovering hidden beauty. Conservators sometimes have to painstakingly remove layers of centuries-old varnish or overpainting to reveal the original masterpiece underneath. It's a delicate, nerve-wracking process – every stroke risks damaging the very thing they're trying to save. Imagine trying to clean a priceless antique teacup while balancing it on the edge of a wobbly table. That's the kind of pressure we're talking about.

The stakes are high because these layers, added by well-meaning hands, often obscure the artist's original vision. Varnishes yellow with age, muting colours that once sang with vibrancy. Overpainting, meant to 'fix' damaged sections, can distort the original work, layering someone else's interpretation over the true masterpiece. Removing these

---

[2] Though the universality of such effects is still under question.

layers isn't just cleaning; it's peeling back time itself, restoring the artist's voice after years of silence.

Take the Sistine Chapel ceiling, for example. For centuries, Michelangelo's frescoes were hidden under a grimy mix of candle soot, dirt, and misguided restoration attempts. When conservators began their work in the 1980s, they faced fierce criticism – what if they over-cleaned? What if the vibrant hues they uncovered weren't 'authentic'? But as the grime was meticulously removed, Michelangelo's true brilliance emerged: skies so blue they seemed infinite, skin tones so luminous they looked alive. The restoration wasn't just about preserving art; it was about reigniting a spark of human creativity that had been dimmed for centuries. This same dance of destruction and preservation happens across the world. In Vermeer's *Girl with a Pearl Earring*, cleaning revealed subtle details – the glint of light in her earring, the texture of her skin – that had been dulled by time. Conservators often describe the process as a dialogue with the artist, a way of collaborating across centuries to let their work shine again.

Here's the tricky part, though: Art restoration is never truly neutral. Every choice a conservator makes – what to remove, what to leave, how far to go – is subjective. It's a blend of science, intuition and ethics. Sometimes, leaving a piece 'imperfect' is the best choice, with the idea that the layers of history should be respected as part of the artwork's journey. After all, even the added 'flaws' tell a story: a story of time, of human hands, of everything that came after the artist's final brushstroke.

And this principle doesn't just apply to paintings or frescoes. Whether it's ancient sculptures, textiles or manuscripts, restoration spans media and eras, always walking the tightrope between preserving the past and respecting its evolution. For instance, when restoring ancient manuscripts, removing water stains or smoke damage can bring clarity, but excessive 'cleaning' risks erasing marginalia or notes added by scribes centuries later – notes that tell us just as much about the artefact's history as the original text.

Ultimately, art restoration is an act of love and trust. It's trusting that the original beauty is still there, waiting to be uncovered. It's loving the process enough to risk damaging what's in front of you, believing that the result will be worth it. And isn't that what all great acts of creation require? A little destruction, carefully wielded, to reveal the masterpiece beneath.

And all of this leads to a profound truth: creation and destruction aren't enemies – they're collaborators. Partners in a dance as old as time. Progress often isn't about adding more but about strategically taking away. Like a sculptor chipping away at stone to reveal a hidden form, or a star exploding to birth new elements, destruction, when controlled, can pave the way for brilliance.

The universe itself is the ultimate proof that creation and destruction are two sides of the same coin. Stars, those glittering celestial factories, spend billions of years forging the elements that make up everything we know. The iron in our blood? It was born in the heart of a dying star, in a final, dramatic act of cosmic creation. When massive stars run out of fuel, they collapse in on themselves and explode as supernovae. These explosions scatter their treasures – iron, carbon, oxygen and more – across the cosmos, laying the building blocks of new worlds.

And here's the kicker: Every atom in our body was once part of a star. The calcium in our bones, the oxygen I'm breathing right now, the carbon that's basically the scaffolding of our entire existence – these were all stardust. We are, quite literally, the universe made conscious. Feeling special yet? I certainly do, especially when I think about how I can barely manage my kitchen, yet the universe managed to turn stardust into life.

Iron's journey is particularly poetic. This element, so crucial to life on Earth, is only made during the most violent of stellar deaths. A star's collapse squeezes atoms together with unimaginable force, creating the dense, resilient metal that eventually finds its way into our blood, our buildings and even into the Iron Pillar of Delhi. The same process that brings stars to their fiery end gives us the tools to live, thrive and create.

So, the next time someone says you're being dramatic, just remind them you're literally stardust forged in cosmic fire. If that doesn't win the argument, I don't know what will.

And the next time something breaks, whether it's your favourite mug or your meticulously planned weekend, take a moment to pause. Maybe what feels like chaos is just creation in disguise. Maybe, like Oreo curling up on the couch after one too many zoomies, the universe is reminding you to take a breath and let the pieces fall into place. Sometimes, breaking is just the beginning.[3]

---

3 But also, sometimes it's just breaking.

7

Stories in the Times of Wi-Fi

As I try to watch a movie, clicking on 'play', that most dreaded of symbols makes its unwelcome appearance: the circle of buffering, spinning endlessly like a digital Ferris wheel going nowhere. It's named the 'throbber' – a word that was probably coined by someone who was frustratedly waiting for content to load. You know it: that hypnotic circle of dots chasing each other in an infinite loop, the digital equivalent of a 'Back in five minutes' sign that nobody believes.

One fun fact is that it started with the Mozilla Firefox logo of a fox chasing its tail around the globe. That spawned an entire genre of circular loading animations because, apparently, we all collectively decided that watching things go round and round somehow makes waiting more bearable. It doesn't.

Oreo, my current critic, shoots me a look of pure feline disdain (yes, he's a dog, but he's very good at channelling his inner cat). His expression practically screams, 'Why can't you just achieve stillness like a normal person so I can continue my sixteenth nap of the day?' He's clearly unimpressed by both my restlessness and the rotating dots that have captured my attention. Meanwhile, the throbber keeps throbbing, in perfect counterpoint to my mounting frustration.

My brain has kicked into troubleshooting mode, running through its well-worn diagnostic checklist like a particularly anxious IT department. Other programs? Dead. Different browsers? Nope. The horrible truth dawns – oh God, the Internet is down! In a moment of pure twenty-first-century reflex, I reach for my phone to Google what's wrong with the Internet, only to be hit with the crushing Catch-22 – no Internet means no Googling to learn why there's no Internet. It's like trying to use a broken flashlight to look for a working flashlight in the dark.

I can feel a looming panic, that special modern strain of anxiety that comes from being disconnected from the global hive mind for more than 0.3 seconds. I take deep breaths. This is alright. After all, the world existed before the Internet. Right?

Although, what did people do before the Internet? Actually, what did they do before television? Before books? Before radio? Before we had all these glowing rectangles and invisible signals keeping us perpetually entertained? Before we all became mesmerized by endless circles of dots chasing each other across our screens?

Oreo, helpful as ever, gives me a look that clearly says, 'I just lie down and sleep, you ridiculous human.' Which, while absolutely on brand for a cat-like dog, and technically sound advice, isn't quite the solution I'm looking for. However, I must admit, his mastery of non-verbal communication is impressive for someone who spends approximately 20 hours a day unconscious.

That's alright, though. I can do something even more primitive, something humans have been doing since we first gathered around fires in caves, trying to make sense of the darkness beyond the flames. Something that predates not just the Internet, but every broadcasting medium we've ever invented. Something that's encoded in our DNA as deeply as Oreo's need to dig holes in inappropriate places.

I'll tell stories.

So ... Here are some stories.

The first is about a glamorous movie star who changed the world of wireless technology forever, though most people didn't know about it until long after she'd left the stage. This is the story of Hedy Lamarr, and trust me, it's not the kind of tale you'd expect from a Hollywood goddess of the 1940s. Then again, the best stories rarely are what you expect them to be ...

It started in Hollywood.

Actually, strike that. The story of Hedy Lamarr begins not in Hollywood but in the elegant salons of pre-War Austria, a world brimming with opulence and shadowed by the impending storm of World War II. Born Hedwig Eva Maria Kiesler in 1914, she was a curious and intelligent child, always drawn to how things worked. Growing up, she had a knack for understanding the mechanics of the world around her, a skill her contemporaries often overlooked.

Her life took a dramatic turn when she married Friedrich Mandl, an Austrian arms manufacturer nearly twice her age. Mandl, a man of wealth and influence, threw lavish parties attended by military experts and politicians, where the latest advancements in weapons technology were discussed over crystal glasses and candelabras. Mandl expected Hedy to be a silent adornment, dazzling his guests with her beauty, but she was anything but ornamental. As the conversations buzzed around her, she quietly absorbed the details of torpedoes, radio controls and military strategies. She listened, learned and filed it all away.

But the marriage was suffocating. Mandl was controlling, limiting her movements and career aspirations. In 1937, she made a daring escape, leaving both her husband and her homeland behind. She fled to Paris, then London and finally crossed the Atlantic to Hollywood. There, under the wing of MGM's Louis B. Mayer, she was reinvented as Hedy Lamarr, the epitome of silver-screen glamour.

Hollywood adored her. Audiences were captivated by her roles in films like *Algiers* (1938) and *Samson and Delilah* (1949). Behind the scenes, however, Hedy's mind was as brilliant as her beauty. She transformed one of her rooms into a makeshift inventor's workshop, filled with tools, sketches and ideas. When she wasn't dazzling the world with her on-screen performances, she was dreaming up innovations ranging from improved traffic lights to ideas for a dissolvable soda tablet.

Then, World War II erupted, and Hedy's past in Austria became more than just a memory. The US Navy faced a dire problem: Their radio-controlled torpedoes, a critical innovation, were easily jammed by enemy forces. The Germans could intercept the radio signals and send the torpedoes careening off course. Hedy, recalling the conversations at Mandl's parties, understood exactly why this was happening.

Her solution was as elegant as it was ingenious. Inspiration struck from an unlikely source: the piano. Watching a pianist's fingers flit across the keys, she thought about how the melody came from deliberate changes in notes, never lingering on a single key. What if, she wondered, a torpedo's signal could hop between radio frequencies in a similar pattern? If both the transmitter and the receiver followed the same prearranged sequence, it would be nearly impossible for anyone to jam the signal because they wouldn't know which frequency to target next.

To bring her idea to life, Hedy turned to an unexpected collaborator: George Antheil, a composer known for his experiments with player pianos.

Not piano players but 'player pianos'. When I was a child and wanted to learn how to play the keyboard, my parents bought me a Casio keyboard, which was more a children's toy than a musical instrument. I never learnt how to play anything more demanding than 'Happy Birthday', but the instrument came with a bunch of pre-recorded tunes – and those I could play without actually pressing a single key. The illusion was never complete though, as no keys moved to match the tune being played.

Why, you may wonder, am I telling you about a toy in the middle of a story about Hedy Lamar and frequency hopping? It's because player pianos – large adult versions of my little electronic keyboard – used to be common instruments in bars in the 1940s.

These could play pre-programmed tunes much like my toy, the key difference being that unlike the pre-recorded digital tape playing on my keyboard, the player piano keys were mechanically moved. The tunes were played using a complex system of vacuum valves. A piano roll, which was a perforated paper roll, was used to encode the tune, and once it was inserted, the piano could play an entire tune independently. Imagine the way a piano starts playing in a ghost movie but in real life, using the magic of valves, levers and ingenuity.

This is the same system that was used for many other early experiments with automations. Paper with punched holes aka perforated paper tape was used in early computing devices ('difference machines') invented by Charles Babbage and in looms used for weaving complex designs on fabrics. I've even seen this in action in Banaras, where some of the traditional weavers still use old looms which work on similar techniques. In fact, such tapes were used for weaving machines before, and were co-opted while building the computing devices.

Together, Hedy Lamarr and George Anthiel developed a system that used perforated rolls of paper similar to those in player pianos to synchronize the frequency changes between a transmitter and receiver. They called it the 'secret communication system'. In 1942, they were awarded US Patent 2,292,387 for the invention.

But the US Navy dismissed the invention. They couldn't believe a glamorous Hollywood actress could devise a serious military innovation. 'Stick to selling war bonds,' they told her. And she did, raising millions to support the war effort. Yet, her groundbreaking idea was left to gather dust.

Many years later, Hedy's brilliance was finally recognized. In the 1950s, engineers revisited her concept, adapting it for military use as electronic technology advanced. By the 1960s, her ideas on frequency-hopping were being used during the

Cuban Missile Crisis to ensure secure communications.

In fact, many foundational ideas of data sharing protocols over wavelengths – like the way we can use code division multiple access (CDMA) and time division multiple access (TDMA) to facilitate wireless communication between multiple devices over the same frequency channel – can find their roots in her work on frequency hopping. Over time, her inventions became the backbone of technologies we now take for granted – such as Wi-Fi, Bluetooth and GPS.

While she was feted as a Hollywood superstar, her contributions to science and engineering went unrecognized for much of her life. Only in her later years did the world begin to appreciate her dual legacy. In 2014, long after she passed away, she was inducted into America's National Inventors Hall of Fame, cementing her place in history not just as a glamorous icon but also as a technological pioneer whose work truly changed the world.

Hedy Lamarr's story is one of resilience and brilliance – a tale of a woman who defied the expectations of her time, proving that beauty and brains are not mutually exclusive. Her legacy reminds us that the most extraordinary minds sometimes shine brightest where we least expect them, and we should not underestimate the quiet person standing in a corner in the party.

Which brings me to the second story, that of the ALOHA[1] network.

---

1 'Aloha' is the Hawaiian greeting that famously means both hello and goodbye. However, ALOHA also has a secret third meaning, as you will discover by continuing to read this book.

It started in Hawaii, late 1960s.

A place where the ocean sparkles, the volcanoes smoulder and the University of Hawaii faces a problem that no amount of sunscreen or good vibes can solve. On the main island of Oahu sits an IBM 360/65 mainframe, a computer so expensive and powerful it might as well be a god in a plastic casing. It's the kind of machine researchers would cross oceans to access – which, unfortunately, they literally had to do.

The university had campuses scattered across the islands, from Maui to Kauai to the Big Island. If you wanted to use the computer, you didn't just log in; you booked a flight. You packed your punch cards (perforated pieces of paper à la player piano), hopped on a plane and hoped your calculations wouldn't crash halfway through.

It was tedious, costly and just plain ridiculous for an institution trying to foster cutting-edge research. Traditional solutions like cables weren't much help. Running wires over mountains and lava fields wasn't going to happen, and undersea cables between islands carried a price tag that could fund a space programme.

Enter Norman Abramson, a professor at the University of Hawaii, who wasn't about to let a little quirk of geography ruin the future of computing. Abramson had a bold idea: Forget wires, what about radio waves? Sure, they were primarily used for communication, but who said they couldn't be used for computers? It was an audacious leap, but it made sense. Hawaii's rugged terrain and scattered islands were perfect for testing a wireless system. Plus, ARPA, the Advanced Research Projects Agency (now DARPA, the Defense Advanced Research Projects Agency), the US government agency funding cutting-edge projects, was intrigued. They were all about pushing boundaries, and Hawaii's logistical nightmare presented an ideal challenge.

Of course, dreaming up a wireless network was one thing; making it work was another. The problem wasn't just about transmitting data – it was also about sharing the airwaves. If two computers tried to talk at the same time, their signals would collide, creating a garbled mess. Traditional networks avoided this by imposing strict schedules or centralized control, but Abramson wasn't interested in playing traffic cop. Instead, he and his team leaned into the messiness of it all.

Their solution was brilliantly counterintuitive: Let the computers transmit whenever they wanted. If two transmissions collided, the receiving computer would simply ignore the nonsense and wait for the senders to try again. The senders, undeterred, would each wait a random amount of time before resending, reducing the likelihood of another collision. It was the computational equivalent of two people talking over each other at a party, realizing it and politely taking turns. It wasn't chaos; it was improvisation.

This method, which became known as the ALOHA protocol, was simple, elegant and revolutionary. By 1971, the first version of ALOHAnet was live. It connected seven computers across four islands and transmitted data at a blazing 9,600 bits per second. Okay, maybe 'blazing' is a stretch, but for its time it was ground-breaking. Researchers could now access the central computer without ever having to leave their own campuses. For the first time, Hawaii's islands weren't just geographically connected; they were digitally linked.

But ALOHAnet wasn't just about Hawaii. It was about

rethinking how networks worked. Before ALOHA, most systems relied on careful scheduling and rigid rules. ALOHA threw those rules out the window, showing that a little randomness and adaptability could go a long way. It was a proof of concept, and this concept of 'random access' didn't just solve Hawaii's problem, it also rewrote the playbook for computer networking.

In 1972, the protocol got an upgrade. The team introduced 'slotted ALOHA', which divided time into neat little slots. Computers could only transmit at the start of a slot, reducing collisions and making the system more efficient. Meanwhile, across the Pacific, a young researcher named Robert Metcalfe was taking notes. Inspired by ALOHA, he adapted its principles to wired networks, creating Ethernet – the backbone of modern Internet connectivity.

ALOHA's influence didn't stop there. Its ideas found their way into wireless systems worldwide, from satellite communications to cellular networks to the Wi-Fi we use today. Every time your smartphone retries a failed message or reconnects to a router, it's channelling a bit of that Hawaiian ingenuity.

What made ALOHA remarkable wasn't just the technology – it was also the mindset. Abramson and his team didn't try to control every variable or eliminate every collision. They embraced the imperfections, proving that a network could thrive not by avoiding chaos but by managing it. It wasn't about forcing order; it was about finding harmony in the unpredictability.

More than 50 years later, ALOHAnet's legacy is everywhere, quietly powering the connections that define modern life. It all started with a problem, an idea and a willingness to experiment.

But, of course, every invention always whispers the name of the one that came before it. To understand ALOHAnet, we need to listen to the echoes of an earlier era, when humans first dared to imagine messages carried invisibly through the air. This story, set in a time before radio, is one that started with another spark of genius that led to communication across oceans.

It started with sparks – literally.

In the late nineteenth century, the world was teetering on the edge of a communications revolution, and no one knew it yet. Telegraphs had already connected continents with wires that carried dots and dashes, but what if you didn't need the wires at all? What if you could send signals invisibly, through the air?

The man who would crack this open was Heinrich Hertz, a German physicist who, in the 1880s, wasn't trying to invent radio. He wasn't even thinking about communication. Hertz was simply exploring the theories of James Clerk Maxwell, who had mathematically predicted the existence of electromagnetic waves nearly two decades earlier. Hertz wanted to prove Maxwell right – or wrong, as the case may be.

In his lab, Hertz built a device that created sparks, generating electromagnetic waves. Across the room, he placed a simple loop of wire with a spark gap between its ends. When the sparks jumped in his transmitter, faint sparks also appeared in the wire loop. It was proof that electromagnetic waves could travel through space and induce currents in distant conductors. Hertz had demonstrated what would later be called 'radio waves'.

But here's the kicker: Hertz didn't think much of it. When asked about the practical applications of his discovery, he famously said, 'It's of no use whatsoever. This is just an experiment that proves Maxwell was right.' Hertz was right, in a way – it wasn't useful *yet*. He died in 1894, never seeing the revolution his work would spark.

Enter Guglielmo Marconi.

Marconi was young, ambitious and – let's face it – not humble. In the 1890s, inspired by Hertz's experiments, Marconi became obsessed with the idea of sending messages through the air. Growing up in Italy, he worked in his attic, cobbling together equipment: coils, wires and rudimentary antennas. He didn't just reproduce Hertz's experiments; he also expanded on them. Marconi realized that by tuning the transmitter and receiver to resonate at the same frequency, he could greatly improve the distance over which signals could travel. He was, in essence, inventing the first practical radio.

At first, Marconi's signals barely travelled beyond his backyard. But he persisted, tweaking and refining, and soon he was transmitting over kilometres. In 1896, Marconi took his technology to England, where he secured a patent for wireless telegraphy. It was the start of something big. With financial backing, he began demonstrating the power of his invention. By 1899, he had transmitted messages across the English Channel, and in 1901, he pulled off the seemingly impossible: sending a signal across the Atlantic Ocean. The letter 'S', tapped out in Morse code, travelled from Cornwall, England, to Newfoundland, Canada – a distance of over 3,000 km.

The world was stunned. No wires, no visible connection – just a message flying invisibly through the ether. Marconi's achievement was a triumph of engineering, but it wasn't without controversy. His work built directly on Hertz's discovery, and many argued that Nikola Tesla, Alexander Popov and others had also contributed significantly to the development of wireless technology.

In fact, the story of one particular scientist needs special mention. It is the one whose name is often omitted from discussions around the invention, just as I did in the previous paragraph, bundling him amongst the measly 'others' when talking about radio and controversy. It is the story of Jagadish Chandra Bose, a polymath who studied everything from electrical signals in plants to electromagnetic wavelengths and data transmission.

You see, Marconi thanked many inventors before him that allowed him to achieve long-distance wireless data transfer. However, missing on the list was the name of an inventor who was then working in the University of Calcutta – a person who had demonstrated wireless signal transmission years before Marconi's celebrated achievement.

The key to Marconi's ability to transmit data wirelessly was a delicate receiver that could detect radio waves and convert them into a physical signal – in this case, a reduced resistance that allowed current to pass through when the radio signals were received. Such a device is known as a coherer.

While the principle was known, there was one big problem. The earliest coherers, developed by physicists like Édouard Branly and Oliver Lodge circa 1890, consisted of metal filings in a glass tube. The filings would 'cohere' (stick together) when radio waves passed through them, temporarily reducing electrical resistance. But the receivers were not as sensitive as required to handle weak signals like the ones Marconi wanted to intercept. They also needed to be reset after every use and therefore could not be used for continuous transmission. Marconi used a receiver that had self-resetting abilities, with liquid mercury instead of metal filings.

But Marconi wasn't the first to do this. Jagadish Chandra Bose had developed an apparatus that did precisely that years before Marconi. In fact, he had demonstrated wireless data transfer during seminars in the Royal Institution in London, and had published the design of his coherer in the journal *Proceedings of the Royal Society*. Science historians have pointed out that Marconi's record of the 'self-resetting mercury coherer' varies widely. He claimed to have gotten the receiver from a friend in the Italian Navy, but many think he created the device based on J.C. Bose's designs.

Regardless, unlike Marconi's commercialization of the technology, Bose wanted to use it for open scientific research. His lack of commercial focus was a significant factor in the diminished recognition for his achievements. He did not initially pursue a patent, and when he did, it was more about establishing scientific priority than commercial protection. While Bose did eventually patent his coherer in 1904, this came many years after his initial demonstrations. By then, Marconi had already secured numerous patents and established commercial operations.

Still, Marconi was definitely the first person to use a Morse-code-based receiver to transmit a transatlantic signal. Eventually, Marconi's persistence and business savvy secured his title as the Father of Radio, even getting him a Nobel Prize, which many scientists believe should have been shared with J.C. Bose.

This tale may be fraught with controversy when it comes to acknowledging scientific contributions, but it is still testament to the ingenuity with which we (meaning the human race) captured invisible waves, converted them to sparks and then turned the sparks into music. In 1906, Reginald Fessenden, a Canadian inventor, made the first radio broadcast, transmitting music and speech instead of Morse code. Listeners on Christmas Eve that year heard Fessenden play 'O Holy Night' on the violin, alongside readings and holiday greetings. It was a new kind of magic: voices and melodies traveling across the airwaves, reaching ears miles away.

Radio didn't just connect continents; it also connected people. By the 1920s, commercial radio stations were popping up around the world, broadcasting news, entertainment and music. Here in India, Vividh Bharati's melodies wove themselves into the fabric of daily life: the morning sounds of All India Radio's distinctive call-sign mixing with the clink of teacups, the evening broadcasts keeping time with dinner preparations, families gathered around their radio sets like earlier generations had gathered around village storytellers. Through the warm static of short-wave came Radio Ceylon's film hits, with children arguing over whose turn it was to adjust the antenna for clearer reception. 'Akashvani' – the voice from the sky – spoke to a nation of hundreds of languages.

Radio became a global phenomenon, transforming everything from communication to culture. And it all started with Hertz's sparks, Marconi's backyard experiments, Bose's ingenuity and creativity in the face of constraints and a belief that invisible waves could carry the weight of human connections.

Which brings me to my next story.

The tale of the invisible waves didn't end with radio of course. Decades after radio's triumphant debut, scientists tried to tune in to something far stranger than human voices – the song of black holes. What they found wasn't what they were looking for, but it would change everything!

It all started with black holes.

Or, more accurately, with the search for them. In the 1970s, Dr John O'Sullivan and his team at the Commonwealth Scientific and Industrial Research Organisation (CSIRO) in Australia, set out to find the faint, elusive radio signals that Stephen Hawking had theorized would be produced by evaporating black holes. The signals were theoretical, hypothetical, and – for all intents and purposes – impossible to detect. But that didn't stop them.

The task was monumental. These signals, if they existed, would be unimaginably faint, scattered across the cosmos and warped by time and space. Radio waves rarely travel in a straight line; they bounce off objects, refract and scatter, arriving at their destination smeared out and distorted. O'Sullivan's team wasn't discouraged. If these signals were out there, they would find them – or so they believed. To do it, they needed to solve a staggering mathematical problem: reconstructing a clear signal from all that chaos.

They developed sophisticated techniques by building on fast Fourier transforms. Remember that trick my mother taught me about taking a big, complex word and breaking it down in order to understand it better? Fourier transforms, which were invented by the French mathematician Joseph Fourier in the early 1800s, essentially do the same to complex waves. A Fourier transform breaks up a wave into multiple simpler waves.

It is a technique that has been used by scientists to decode MRIs, in automated complex image processing to understand quantum physics, by audio engineers to decode and clean up complex sound waves, by economists to analyse market trends and by mathematicians for all sorts of things.

While extremely useful, traditional Fourier transforms can be time consuming. Imagine if you had to break down every word that you read in a book – it is possible, but the process would be slow. The team at CSIRO used a variation of Fourier transforms called fast Fourier transforms (FFT).

FFTs use mathematical shortcuts like 'Hamming windows' to make this process much faster. To extend the analogy I used before, instead of having to break down every single word in a book, we may choose to focus on a few words and try to decode only these, allowing us to make sense of those specific words much faster. Hamming windows are mathematical constructs that allow us to focus on a specific area of a wave and reduce the disruption at the edges of the area we choose to analyse, making sure that the focus fades out gently, rather than abruptly.

The scientists at CSIRO used FFTs and built on the same principles to break the signals from the edge of time down into their components and then pieced these back together like an astronomical jigsaw puzzle.

The team worked for years. But the signals didn't come. The

black holes, if they were there at all and evaporating, remained stubbornly silent. The experiment was, in the strictest sense, a failure. But in the process, the team had created something extraordinary: a method for untangling distorted radio waves, and while this tool seemed to be useless for detecting evaporating black holes, it had enormous utility in several less esoteric tasks.

Fast forward to the 1990s, and O'Sullivan's forgotten tools were about to become very relevant. Wireless communication was on the rise, and with it came a massive problem: Indoor wireless signals were a mess. Radio waves inside buildings didn't behave neatly. They bounced off walls, floors, ceilings and furniture, arriving at receivers in fragments, delayed and distorted – a phenomenon called 'multipath interference'. At high data rates, this interference turned radio signals into indecipherable soup.

The problem felt eerily familiar. O'Sullivan recognized that the mathematical tools he and his team had developed to hunt for black holes decades earlier could be adapted to solve the problem of multipath interference. He strung together a team that included David Skellern and Terry Percival, and they refined the techniques, combining them with new ways of encoding data across multiple frequencies and managing interference. What emerged was something amazing: a way to make wireless communication reliable, even in chaotic environments filled with static and electronic noise.

In 1996, CSIRO was granted Patent 5,487,069 – a patent that would go on to underpin Wi-Fi technology as we know it. The work wasn't flashy. It wasn't heralded as a revolution at the time. But it solved one of the most stubborn challenges in wireless networking: making signals reliable indoors. And it started with the 'failure' to find black holes.

What makes this story remarkable isn't just the technology – it's what it says about pure research. O'Sullivan's team wasn't trying to invent Wi-Fi. They weren't trying to change how the world communicated. They were chasing a theoretical idea about the universe, a question about the nature of black holes that still hasn't been answered today.

Their work was driven by curiosity and not by a quest for commercial application. When it didn't yield the results they hoped for, it might have been easy to call it a failure and move on. But science doesn't work like that. The tools they built for one purpose found utility in another cause. An apparent failure turned out to be something very valuable.

Wi-Fi didn't spring into existence overnight; nor was it the work of one team alone. The CSIRO innovation solved a critical problem – indoor signal reliability – but the final product required contributions from researchers worldwide tackling everything from protocols to hardware. Still, O'Sullivan's work was pivotal, so much so that major tech companies like Microsoft, HP and Dell eventually paid licensing fees to use the technology. CSIRO defended its patent in court and secured hundreds of millions of dollars, money that went back into Australian research[2] to fuel the next generation of discoveries.

In 2012, O'Sullivan and his team received the European Inventor Award for their work on Wi-Fi. But the award wasn't

---

[2] CSIRO used the funds to establish the Science and Industry Endowment Fund (SIEF).

just for their technical achievement – it was for what their story represents. They weren't looking to revolutionize wireless communication. They were trying to hear the whispers of dying black holes, to glimpse something no one had ever seen. They didn't find it. But in their failure, they built something extraordinary.

Pure research doesn't always give you the answers you're looking for. Sometimes, it doesn't give you answers at all. But every so often, it leads you somewhere unexpected and wonderful. O'Sullivan's team didn't find black holes, but they found a way to make the world more connected, one signal at a time. Their 'failure' is a reminder that in science, even when you miss the mark, you might just hit something else entirely.

Which brings me to a personal story – that of my failed watercolour painting adventures.

It started, as most things do, out of sheer necessity. I wasn't trying to make masterpieces or change the way I saw the world. I just needed something to do – something to pass the time and quell my anxious mind. Painting seemed like a good option. It didn't demand too much from me. It wasn't a competition. Just colours, a brush and a little bit of focus to keep the spiralling in my head at bay. And I had paid for the colours already.

So I started.

But here's the thing about me. When I get obsessed with something, I dive in headfirst and keep going, no matter what hurdles come my way. It's never just a hobby – it's a full-blown commitment. When I first picked up painting, it was supposed to be a way to pass the time and calm my anxiety. But the moment I dipped that brush into paint and saw colour come to life on the canvas, I was hooked. Suddenly, I wasn't just painting to relax; I was painting to *create*. Every stroke felt like a small step toward something bigger, something I couldn't quite define but needed to chase.

Then I sprained my thumb. A ridiculous, clumsy injury, the kind that should've been a sign to pause and take a break. Instead, it felt like a personal affront. Painting had become my escape, and I wasn't about to let a sore thumb take that away. So, I did what any obsessed person would do: I adapted. I picked up the brush with my left hand, told myself it was temporary and braced for chaos.

And chaos it was. At first, my left hand felt like it belonged to someone else – clumsy, uncooperative, like a toddler trying to hold a spoon for the first time. The branches I tried to paint wobbled unpredictably, dipping too far one moment and stopping short the next. My usual neat arcs and precise strokes (at least more precise than with my 'stupid hand') were nowhere to be found. I nearly gave up on that first attempt, but something kept me going. Maybe it was the obsession, or maybe it was the tiny voice in my head whispering, 'Just see where this goes.'

So, I kept painting, and somewhere in the mess, I started to notice something. Those wobbly branches didn't look wrong – they looked real. The curves and dips weren't stiff or forced; they had a life of their own. They bent and twisted like they'd grown that way, shaped by years of wind and gravity. For the first time, my trees didn't look like something I'd planned but something that had *happened*.

That's when it clicked. The so-called 'mistakes' my left hand made weren't mistakes at all. They were possibilities.

They were my brain letting go of control and my hand finding its own rhythm. The process became fun in a way I hadn't expected. Every wobble, every smudge, every uneven stroke added something new, something surprising. My art stopped being about what I thought it *should* look like and became about what it *could* look like.

Even after my thumb healed, I couldn't go back to painting the way I used to. My right hand still had its place, but my left hand became my co-conspirator, my partner in embracing the unpredictable. I started leaving room for the unexpected – letting paint drip, smudging a corner on purpose or switching hands mid-stroke just to see what would happen. It wasn't just painting anymore. It was a conversation between me and the canvas, where the 'mistakes' were as much a part of the story as the deliberate strokes.

This taught me that sometimes you get the best results when you stop trying to control everything and let a little chaos in. Whether it's a left-handed branch that grows its own way or a network that embraces collisions, the magic happens in the imperfections. It's messy, unpredictable and absolutely worth it.

Just like my left hand discovered a new way to paint by embracing its limitations, each breakthrough in wireless technology came from embracing rather than fighting chaos. The ALOHAnet engineers didn't try to prevent signal collisions – they accepted them as inevitable and built a system that could recover and retry. O'Sullivan's team didn't fight against the mess of scattered radio waves; they developed mathematical tools to untangle them. And Hedy Lamarr didn't try to stop enemies from intercepting signals – she made the signals dance unpredictably between frequencies, turning randomness into security.

And here's where I confess my delightfully devious plot device: While you were wandering through tales of Hollywood glamour and Hawaiian chaos, of German sparks and cosmic quests – and yes, even my own stumbling adventures with a paintbrush – these stories have been quietly assembling themselves into something larger, like radio waves finding their way through the chaos.

While reading through this meandering set of anecdotes, maybe without even realizing it, you've absorbed a few bright threads in the vast tapestry of how modern Wi-Fi works. These stories are threads in a tapestry that may look simple at first glance, but the more closely you look, you see how each thread makes its own contribution to the whole.

There are hundreds of brilliant minds in labs and garages, teams and individuals working through the night to solve just one more problem. Each has stories of failed experiments, and some of those failures led to unexpected breakthroughs. And all these stories are transmitted through the oldest protocol we have: storytelling.

Think about it. There was the all-too-human drama of the beautiful Hedy Lamarr's spectacular escape from her controlling arms-dealer husband. But in the telling of that story, we also learned about the concepts of frequency hopping and secure communications.

When you chuckled at those ALOHAnet researchers saying 'Eh, let the signals crash, we'll try again', you were also stealthily downloading the idea of random-access protocols into your consciousness.

That seemingly random story about O'Sullivan's team failing to find black holes? That sneakily installed the concepts of signal processing knowledge and the mathematics of radio wave untangling in your neural pathways. While you thought you were just enjoying a tale of cosmic disappointment turned technological triumph, you were also in fact absorbing the foundational principles of how your favourite web series reaches your screen of choice (when it's not buffering!).

It boggles my mind when I think about it. Every time I pull out my phone to check social media or stream a show, I'm witnessing a remarkable collaboration across time. My device is doing the digital equivalent of that polite party shuffle – backing off and retrying when it bumps into other signals, just like ALOHAnet taught us. It's also playing a high-speed game of musical chairs with frequencies, bouncing faster than I can blink, using Hedy Lamarr's clever idea to keep my data secure.

The data in question is being blasted into space, with us using the same concepts originally developed to hear and intercept signals from deep space. However, now we use satellites that intercept and send these signals back to earth. Those satellites used to broadcast content to millions of (analogue) TVs, allowing someone in a remote village to learn the cricket score. Now, the satellites use the same bandwidths to send different types of content to billions of digital devices. By using advanced techniques like CDMA and TDMA, which trace their roots to frequency hopping concepts pioneered by Hedy Lamarr, these satellites allow someone in a city to Google how much coffee is too much coffee[3]. In a rural small town, these satellites can send data to a young student studying Fourier transforms through an online video. She may be unaware that the same mathematical concept is at work to deliver the clear, stutter-free video she's watching. And here, on my couch, the satellite sends data packets that allow me to work on a cloud-based text editor while I simultaneously track how many runs my favourite cricketer scored.

What's even more fascinating is that those same radio waves that Hertz sparked to life in his lab and Marconi sent across oceans are still the backbone of our wireless world; only, they're singing at different pitches. That unassuming Wi-Fi router in the corner of my room orchestrates a complex dance across multiple frequencies – 2.4 GHz for reaching those tricky corners of my home, 5 GHz when I need extra speed for streaming and, in newer devices, 6 GHz for when I *really* need to move data fast. Like a conductor leading different sections of an orchestra, my router juggles these frequencies, each with its own strengths and quirks. The lower frequencies punch through walls better, while the higher ones move data faster but prefer clear lines of sight.

When these signals start pinballing around my apartment, bouncing off walls and ricocheting off furniture, my device uses mathematical tools – originally designed for listening to black holes – to make sense of the chaos. Add in modern touches like multiple antennae working in harmony and beamforming (imagine a spotlight focusing its beam rather than lighting up the whole room), and you've got a system

---

[3] The answer obviously is, 'There's no such thing as too much coffee.'

that would make those early pioneers' heads spin.

Together, these solutions turn what should be a complete mess – like trying to hear a whispered conversation in a crowded room – into the magic of modern connectivity. The Internet is, by some estimates, a place where over 3 billion of us want to watch, do and learn about everything, everywhere, all at once. But of course, sometimes that spinning circle shows up, reminding us that even magic has its limits.

So, while Oreo sleeps blissfully, you've downloaded the blueprints of that spinning circle on my screen. From Hertz's 'useless' sparks to CSIRO's very profitable patents, from Lamarr's piano-inspired frequency dancing to ALOHAnet's embrace of chaos – it's all there in your head, nestled comfortably between your favourite movie quotes and your shopping lists.

Who knew science could be so sneaky?

Maybe that's the real revelation: In the end, science isn't just formulas and patents. It's about humans, making magnificent mistakes, having absurd ideas and, occasionally, revolutionizing the world accidentally while looking for something else. Maybe someday we can talk about the time a melted chocolate bar ended up inventing the microwave?

8

Light and All It Touches

The animated movie finally loads after I reset the Wi-Fi by using the nuclear option – turning it off and on again. I'm braced to dive deep into the world of this beautiful animated movie, and I turn off all the lights and settle on the couch. Now, I don't know about you but as far as I'm concerned, there is a direct connection between sitting still to watch a movie and an immediate attack of the munchies. I had anticipated the request from my tongue and kept a pack of chips on the table. Feeling very impressed with my planning skills, I reach for the chips only to knock them over in the dark.

In the blue glow of the screen, I scramble to clean up, muttering apologies to a very unimpressed Oreo. But as I pick up the bag, something strikes me. The simple electronic light coming from the TV screen has transformed my familiar living room into a landscape of shadows and highlights with the blues and the yellows reflecting off the shiny surfaces, and the darks and shadows becoming deeper turning the mundane into something almost magical. The black patches of Oreo's fur become even blacker, as he quietly settles down to snooze, immediately falling asleep in a way that never fails to amaze me or make me jealous.

It makes sense that I am in this poetic mood. You see, I am watching a Studio Ghibli movie. As everyone who's been following the state of the art in deepfakes knows, Ghibli is a Japanese studio that makes beautiful animated movies with a wonderful watercolour gouache-like quality. To me, these movies feel like visual poetry – plus there's the music and, of course, the stories. Every frame is hand painted, lending an ethereal quality to the work.

In fact, something I love doing is to pause the movie when an exceptionally beautiful frame comes along, and try to make a still painting out of it (using watercolours, not AI). My work is easy - I just have to copy what's in front of me; yet at the same time incredibly difficult since every frame is a classic and the beauty in it has so many layers to it. As I pause to admire a frame, I am struck by the use of light. The way the Studio Ghibli artists use light to evoke wonder and add dimensions is breathtaking, no matter how many times I see it.

Come and join me as we look at the light in this paused frame.

In the scene before me, light doesn't just illuminate – it reveals. It speaks in ways that don't need words, shaping how we feel before we even understand why. A golden sunset isn't just the end of the day; it's the soft warmth of nostalgia, the slow exhale of something coming to a close, the quiet promise of tomorrow.

A single lantern flickering in the dark isn't just a source of light – it carries the weight of solitude, the hush of something sacred or the quiet reassurance that even in darkness, you are not completely lost. And then there are the god rays – those beams of light breaking through clouds, through trees, through the atmosphere itself, as if the sky is *breathing*. They don't just highlight a space; they transform it, making the ordinary feel divine.

It also makes me *feel*. It's a special kind of magic. By just manipulating the light in this scene, artists have been able to invite me into an entire universe. Light adds depth where there was only space, carving out shape, distance and movement even in stillness. It shifts, bends and glows. Light shining through mist makes the air thick with possibility. The shimmer of sunlight on water makes a lake feel restless, alive. That's why a single frame, bathed in the right light, can feel like a memory you've never lived, a story you were always meant to know.

Just like in this frame, in the world around us as well, light is magical. It is something that affects so many aspects of our life, and yet we take it for granted. We humans are a very visual species, relying on changes in light to recognize the changes around our environment. Our visual system – the part of our body that helps us sense and understand the visual cues from our environment – evolved to be exquisitely sensitive to motion and changes in luminance. Deep in our visual cortex, specialized neurons fire rapidly when they detect changes in brightness and movement, a trait that helped our ancestors spot both predator and prey even in dim lighting.

This motion detection system is so fundamental that it kicks in before our conscious brain even processes what we're seeing. This is also why if I'm staring at a tree, I may not be able to see a bird sitting on a tree branch, but the moment it takes flight I spot it.

This is why animation is so powerful; it can create such compelling illusions with just simple shifts in lights and shadows. When early animation pioneers discovered they could suggest complete forms and fluid movement with nothing but contrasting tones, they weren't inventing something new. They were tapping into one of our oldest visual capabilities. Our brains are so primed to construct meaning from motion that we can perceive full, complex scenes from just a few strategic shifts in brightness.

Weirdly enough, with such forms, less is more. Early digital animators learnt this the hard way. With advances in computing, they realized that adding more complexity to their characters had the opposite effect to that intended. Instead of making the characters feel more lifelike, it made them veer into what is now known as the 'uncanny valley'. When there are too many details, our brain focuses on the cues that are missing, making the characters in front of us feel alien and absurd, and not in a good way.

Contrast this to the technique Ghibli is using in front of me. Here, the artists have used broad strokes and large changes in contrast to convey the scene. Since our brain is using these cues to fill in the rest, the end result looks far more convincing to us than if the artist was too focused on adding every single detail. It is also prudent to remember that these animated movies aren't the first media to rely on our ability to construct whole premises with light and shadow for storytelling. Artists and storytellers from many cultures from around the world have been using similar techniques for the longest time; as with many things, we use principles shown to us from generations ago. Our species' connection with art extends all the way back to our ancestors using charcoal and other pigments to paint on cave walls.

A memory hits me suddenly. It's from my PhD years, while I was taking a break from the endless hours of staring at code that refused to work. That day was different from my usual days, as there was a special performance happening on campus. Traditional shadow puppet artists were going to do a show and everyone in the university from students, professors and admin staff were lining up outside Dasheri to get a seat for the show. (The auditoriums in the National Centre for Biological Sciences are named after mangoes. There are also Malgova, Safeda, and Langra, among others.)

The evening air carried the scent of coffee and samosas from stalls outside the auditorium as I found myself drawn to a shadow puppet performance. After standing in the line for a long time, I found a seat and sat in front of the stage, just as the performance began. The leather puppets danced behind a white screen, their shadows transformed by carefully positioned lights into living, breathing characters.

What captivated me wasn't just the visual magic – even though the visual elements in front of me were nothing short of breathtaking – it was how the entire performance came together. The puppeteer's voice shifted effortlessly between characters, from the resonant declarations of Ram to the melodic responses of Sita. Traditional songs, accompanied by percussive notes and musical instruments, were woven through the narrative, their rhythms matching the puppets' movements perfectly. Even though I could see the mechanics – the sticks controlling the puppets, the lights creating the shadows, the artists singing at the side – my brain happily surrendered to the illusion. The scientist in me was analysing this willing suspension of disbelief but, in that moment, it felt more like magic than neuroscience.

But, like with most magic, thinking rationally about the other elements of this performance only added to it. The puppeteers' mastery became even more impressive when I understood the sheer complexity of what they were doing. Each puppet was an intricate work of art, crafted from treated leather with delicate perforations and joints that allow for fluid movement.

The puppet's translucent surface created subtle gradients when light passed through – thicker areas appeared darker, while the intricate cutwork produced lighter patterns. A single character might have had multiple movable parts, each controlled by thin bamboo sticks, requiring the puppeteer to orchestrate a complex dance of shadows with just their hands.

The puppets were made of leather aged in a very specific way. This gave them an almost translucent quality, with the light seeping through them differing based on how close to the light source they are. The subtle paint interacts with the light, making the areas painted as black eyes look darker in a shadow. The artist was not only keeping the beat, telling the story and moving the different parts but also moving the puppets forward and backward to change the shadows' sizes, making certain characters smaller or bigger depending on the scene.

What's remarkable is that while I didn't understand the language being sung, the performance transcended language. When a character's head tilted in grief or a hand gestured in defiance, I understood the body language perfectly. The universality of these movements, combined with the music's emotional resonance, meant the story reached across linguistic barriers. The brain, it seems, speaks the language of movement and emotion fluently, even when words escape us.

This marriage of light and shadow to tell stories isn't unique to India. Across Asia, similar traditions evolved independently and each had their own fascinating twists. In Indonesia, the *wayang kulit* tradition uses elaborate puppets to perform episodes from the Ramayana and Mahabharata, but with local interpretations and additions. Chinese shadow theatre, *píyǐngxì*, often features puppets with translucent, coloured parts that add another dimension to the storytelling. In Turkey, the Karagöz theatre developed its own distinct style, often using humour to comment on social issues.

What connects these traditions isn't just their use of shadows, but their understanding of human perception. They all recognized that our brains are hardwired to find patterns, to construct narratives from fragments, to see life in the play of light and dark. A slight tilt of a puppet's head, a tremor in the puppeteer's voice, a subtle shift in the music – and suddenly we're fully invested in the fate of these shadow beings.

I experienced this myself that evening. Though I couldn't follow the words, I found myself holding my breath when an overbearing Ravana appeared on screen, his multiple heads and arms creating an imposing silhouette. I understood the flying Hanuman's loyalty through the way his shadow bounded across the screen, the puppeteer's movements giving him a sense of boundless energy and devotion. When Sita's delicate silhouette moved gracefully across the screen, her gestures spoke of both dignity and despair, transcending the need for verbal explanation.

As I look at the paused frame in front of me, I feel similar emotions to what I felt that day during the shadow puppet performance. Ghibli frames radiate a sense of whimsy and joy, whether through exaggerated shapes, lively movement or small, unexpected details that make the world feel alive. What's fascinating is how they achieve this through careful manipulation of light and motion, even in seemingly static scenes.

Take *My Neighbor Totoro*'s forest scenes, for instance. This is one of my favourite Studio Ghibli movies. I heard in an interview once that while adapting the movie for the US audiences, the studio considered changing the name of the character from Totoro to Dave. Thankfully, they didn't. I cannot imagine *My Neighbor Dave* having the same whimsy as *My Neighbor Totoro*.

In any given frame, I can notice dust motes dancing in sunbeams, leaves trembling in barely perceptible breeze and shadows that shift and play across surfaces. These minute movements – what animators call 'secondary animation' – create a world that feels perpetually alive. Even when the main characters are still, the background breathes.

The connection hits me – whether it's a leather puppet creating emotions through shadows, or a hyper-realistic drawing emerging from careful gradients, they're both playing with the same fundamental principle.

Take the Indian miniature paintings I grew up seeing in history books and trying to replicate with my limited tools and training. While known for their vibrant colours, what gives them their power is their mastery of tone. Look closely at a Mughal miniature, and you'll see how the artists used subtle variations in value (different intensity of the same shade of colour) to model form, to make figures appear three-dimensional on a flat surface. They understood that our brains interpret changes in brightness as changes in depth, a trait that probably evolved from spotting predators in dappled forest light. Though, as frequently happens across human history, it isn't a conscious learning of underlying neuroscience principles but an intuitive understanding of shapes and forms in the world around us.

The old Western painting masters weren't doing anything fundamentally different from ancient Indian artists. When Rembrandt painted his self-portraits, he wasn't just showing off his ability to capture his own likeness – he was also conducting experiments in light and shadow. The famous 'Rembrandt lighting' technique, where a triangle of light appears under the subject's eye, isn't just artistic flourish. It's a sophisticated understanding of how our visual system interprets form through the interplay of light and dark.

This signature triangle – sometimes also called 'Rembrandt's pearl' – appears on the shadowed side of the face, creating a small, bright patch roughly the size and shape of the subject's eye. It results from light wrapping around the nose and catching the cheek while the rest of that side of the face remains in shadow. But what makes this technique so powerful isn't just its distinctive look – it's how perfectly it aligns with the way our brains process faces and form.

You see, our visual system evolved to be extremely sensitive to faces, particularly the area around the eyes. Therefore, when we look at a face, we instinctively search for key points of light and shadow that help us understand its three-dimensional structure. The Rembrandt triangle provides crucial depth information exactly where our brain most actively seeks it. It creates a perfect balance – enough shadow to suggest form, enough light to reveal detail.

This wasn't just artistic intuition on Rembrandt's part, though I'm sure intuition played a huge role. He experimented extensively with light, often using his own face as a subject. His self-portraits show a deep understanding of how different lighting angles affect our perception of depth and character. He discovered that by placing the main light source high and roughly at a 45-degree angle to the subject, with the shadowed side of the face receiving just enough reflected light to create that triangle, he could achieve maximum dimensional effect with minimal means.

The technique is so effective that it became a standard not just in painting, but in photography and film. Early Hollywood cinematographers, working in black and white, relied heavily on Rembrandt lighting insights to create depth and drama. Even today, portrait photographers use this 400-year-old technique because it simply works – it triggers our brain's built-in ability to construct three-dimensional information from patterns of light and shadow.

What's particularly fascinating is how this technique works across different faces and features. The triangle's size and exact placement might vary, but as long as it maintains certain proportions – roughly eye-sized, touching the eye line, bordered by the nose shadow – our brains recognize it as a marker of three-dimensional form. It's like Rembrandt discovered a universal language of light that our visual cortex inherently understands.

Modern digital artists, armed with tablets and software, still rely on these ancient concepts. I've spent embarrassing amounts of time watching speed-paints where artists start with simple values – basic light and shadow – before adding any colour because they know, just like the shadow puppeteers and the old masters, that if you get the values right, the brain fills in the rest.

Art, especially paintings, has always affected me viscerally. I do not claim to be unique in that aspect. Many of us instinctively have reactions to art that we can't quite comprehend. We also seem to have a sense of whether something is an 'original' artwork or not. Everyday humans – and not just art experts – seem to be able to tell that something is 'off' when we look at a replica of a painting vis-à-vis the original, though the change in perception also happens if someone just tell you that a painting is fake. So while instinct plays a part, so does our preconceived notions.

Regardless, there is something indescribable that happens to me when I'm left alone around paintings. I'm always that person in museums – the one who parks themselves in front of a single painting while other visitors flow around them like a river around a stubborn rock. My friends have long since learned to either lose me entirely or drag me along when their patience runs out. But this one time was different …

It was my birthday. I had gone to Europe on a scholarship to present my research in a conference, and for my birthday, I chose to tick off one of the biggest items on my bucket list – visiting the Louvre Museum in Paris. While most people were queuing up to look at the famous *Mona Lisa*, I was standing transfixed in a different section. Here, I was standing in front of a large painting – large enough that I stood side by side with the characters in the frame. Even for me, known for spending absurd amounts of time with individual artworks, this was something else. The world didn't just quiet down; it vanished entirely. The painting had pulled me in so completely that I stood there, utterly motionless, lost in how the artist had captured light cascading across the canvas.

Until, that is, a small child crashed into my legs at full speed, followed by their mortified mother's 'Oh my god, I'm so sorry! We thought you were part of the exhibition!' In hindsight, I could hardly blame them – there I was, frozen in place wearing a flowing white Grecian looking dress, standing completely still in front of a classical painting. However, while it is a cool idea to use a living person to turn a painting into an installation, I was also sporting a bag containing a phone and camera and wearing old, red canvas shoes, which should've been a dead giveaway that I was not in fact part of the exhibit.

The initial embarrassment of the collision quickly gave way to an unexpected warmth. There was something oddly delightful about being mistaken for art, even if it was just because I'd been standing unnaturally still in an appropriate costume. For a brief moment, I had been part of the museum's magic – not just an observer but a piece of the exhibit itself, my own draped fabric and posed stillness becoming part of the play of light and shadow I'd been so absorbed in studying.

The mother was still apologizing, but I found myself smiling. In a way, hadn't the art done exactly what it was supposed to do? It had absorbed me so completely that I'd transformed from viewer to unwitting installation piece, the boundary between art and audience dissolving in that moment of perfect immersion. Perhaps that's what great art does – it doesn't just invite us to look, it also invites us to become part of its world, even if that sometimes means being mistaken for a statue.

As I stood there, inadvertently impersonating classical sculpture in my white dress, I couldn't help but think about the irony. Most people, myself included, grew up imagining ancient Greek and Roman statues as pristine white marble, the epitome of classical restraint. I remember the first time I read that these statues were actually painted in vibrant colours – I straight up didn't believe it. It seemed impossible, almost sacrilegious to imagine the *Venus de Milo* in anything but pure white.

But the evidence is overwhelming. Using advanced techniques like ultraviolet light and X-ray spectroscopy, archaeologists have detected traces of pigments on classical sculptures. The term for this ancient painted statuary is 'polychromy', and it turns what we think we know about classical art on its head. The *Augustus of Prima Porta*, one of the most famous Roman statues, wasn't a stark white figure – it was painted in rich colours, with red and purple robes, flesh-toned skin and even brown hair. This is true for a lot of ancient Indian temples as well.

The paint simply didn't survive millennia. Most ancient pigments were organic materials that degraded over time, leaving behind the bare marble we see today. It's a bit like finding a faded billboard from the 1950s and assuming it was meant to be sepia. This misconception about classical statuary was so deeply ingrained that when evidence of the original colours was first discovered in the nineteenth century, many scholars refused to accept it. The idea of pure white classical sculpture had become fundamental to Western art history – it influenced Renaissance artists, neoclassical sculptors and even modern architecture with its beige aesthetic.

The truth is, these statues were designed with a sophisticated understanding of how paint and sculpture work together. The artisans carved subtle details into the marble knowing they would be enhanced by colour. The eyes, for instance, were often painted in incredible detail, turning blank stone into startlingly lifelike gazes.

Speaking of painted statues and skin tones, there's something remarkable about how ancient artists managed to capture the translucent quality of human skin – a feat that relies on understanding how light actually interacts with our skin. Let's go back to the oil painters. They would start with a base layer of paint, often with greenish or greyish undertones (called verdaccio) – that is, they painted the whole canvass grey or green.

This is a technique many painters use, and on first glance, it seems counterintuitive. Surely painting a colour on top of a colour is harder than painting on a blank white canvas. When I was trying to learn oil painting, I found this extra step both boring and unnecessary. However, my paintings would just not look like what I'd wanted, with the end result being me using too much of either the darker or lighter colours that just didn't *look* right.

Our brain does not interpret every pixel or every colour in front of us as an independent unit. It looks at the whole picture, making the same pigment appear darker or lighter based on what is placed next to it. This is because of colour theory, or simultaneous contrast. This is also why the exact same shade of makeup products like lipstick looks wildly different on people with different skin tones, or why sometimes a grey shirt looks darker when paired with white trousers as opposed to black or dark brown trousers.

This is important because when we look at a blank canvas, the white of the background is too light, causing artists to pick colours that are too dark for the darker areas and too light for the lighter areas. Underpainting – the technique of first painting the blank canvas in a mid-toned grey, green, orange or pink – allows us to naturally be able to see the contrasts between pigments when we put them down on the canvas, thereby helping create a more believable colour palette.

The next thing these artists did was called glazing – building up multiple thin translucent layers of colour, instead of painting on the final colour directly. By doing so, they were actually manipulating how light interacts with the whole painting ever so subtly. Each layer they painted would interact with light slightly differently, creating that ineffable quality of living flesh, giving that luminous quality that makes their portraits glow. It's remarkably similar to how actual skin works, though they figured this out through observation and intuition rather than by reading medical journals.

When you look at Vermeer's *Girl with a Pearl Earring*, you're not just seeing paint on canvas – you're seeing light performing an intricate ballet through dozens of near-transparent layers. The effect is so convincing that her skin seems to contain the same depth and luminosity as real human skin. It's the seventeenth-century equivalent of high-definition rendering, achieved with nothing but pigments, oil and an understanding of how light behaves that was centuries ahead of its time.

These artists, again, weren't just copying what they saw on the surface but recreating the actual optical processes that make skin look like skin without knowing the physics behind it. Today, we know that skin isn't just a surface – it's more like a fog bank that light gets lost in. What happens is a process known as subsurface scattering. When light hits our skin, it doesn't bounce off like it would from a mirror. Instead, it performs a complex dance, penetrating different layers, bouncing around inside and eventually finding its way back out. Different layers of skin and the muscle underneath absorb some wavelengths of light and reflect others to varying degrees, thereby making our skin look 'living'. This is why we don't look like walking mannequins – even if the shades those mannequins are painted are exactly the colour of someone's skin.

What is remarkable is that for both these processes, the real magic happens at the microscopic level, where light reveals its peculiar dual nature. It's both a particle and a wave, which is about as intuitive as Oreo being both awake and asleep at the same time. Yet this duality is fundamental to how everything works, from Rembrandt's glazing techniques to the device you might be reading this on.

When light hits those translucent layers of paint, it does two things simultaneously. As a wave, light bends and interferes with itself as it passes through each layer, creating subtle variations in colour and intensity. As a particle –

photons – it bounces around like billions of tiny ping pong balls, some getting absorbed, others emerging changed. This quantum dance creates that quality that makes painted skin look real, even though we're just looking at pigments suspended in oil.

The old artists were quantum physicists without knowing it, intuiting through countless hours of observation how to manipulate light's dual nature.

Here's the even more fascinating bit: This same duality that makes the *Mona Lisa*'s skin glow is what we now harness to etch microscopic patterns on silicon chips. In photolithography, we use light's wave nature to create interference patterns smaller than the wavelength of light itself – a technique that would have blown Vermeer's mind.

Computer chips are microscopic light paintings, and these works of miniaturized art form the brains of our electronic devices. Each transistor is a sculpture carved by light, each circuit a pathway etched by carefully controlled waves of photons. In a way, we're still artists like Vermeer and Rembrandt, but instead of painting faces, we're painting the pathways that carry our thoughts, our messages and our connections to each other.

The journey of how we have grown to use light from art to technology doesn't end in our silicon chips. Perhaps the most beautiful transformation has been in how we've learnt to trap light and use it to carry our messages across vast distances. In fibre optic cables stretching across ocean floors, light performs an endless dance, bouncing from wall to wall inside glass strands thinner than a human hair.

Remember how light bounces around inside our skin

before re-emerging? Well, imagine if you could trap that light and make it bounce along a path instead of scattering everywhere. That's exactly what fibre optic cables do, using a principle called total internal reflection. It's like creating a water slide for light – once it's in, it has no choice but to follow the tube's path, bouncing off the walls at just the right angle to keep zooming forward.

Picture this: You're in a swimming pool (or remember being in one), and you shine a flashlight up at the surface from below the water. At certain angles, instead of the light passing through to the air above, it bounces back into the water. That's total internal reflection, and it's the same principle that keeps light signals trapped inside fibre optic cables, carrying our cat videos and work emails across oceans at literally the speed of light.

Think about that for a moment – when you send a message to someone on the other side of the world, your words become pulses of light, racing through crystalline threads in the darkest depths of our oceans. The same physics that makes a crystal goblet sparkle and gives diamonds their fire now carries our posts and cat videos across continents. There's something poetically fitting about using light to illuminate the darkest parts of our planet.

Inside these glass threads, light reveals another aspect of its nature – its purity. Unlike electrical signals that degrade and need constant boosting, light can travel enormous distances while maintaining its integrity. The light bounces along the fibre's walls at precisely the right angle, like a cosmic game of billiards played at the speed of, well, light. Each photon follows this predetermined path, carrying bits of our digital lives through the oceanic abyss.

These undersea light paths form a global nervous system, pulsing with information day and night. Every time you stream a video or send a message, you're actually sending light on a journey through this vast network. Your data becomes a stream of photons, dancing through glass threads, under mountains and oceans, emerging somewhere on the other side of the fibre optic cable to reconstruct your words, your images, your connections.

But there's an unexpected threat to this global communication network: sharks. Yes, sharks. These magnificent predators, for reasons that continue to puzzle marine biologists, seem to have a particular fondness for trying to bite through submarine cables. One theory is that they mistake the electromagnetic fields around the cables for bioelectric fields produced by prey (sharks can detect electromagnetic fields, which is a sense we lack). Or maybe, they're just really committed to disrupting our Zoom meetings.

The problem was serious enough that cable companies had to start developing shark-resistant armour for their cables. Think about that for a moment – somewhere out there, there's an engineer whose job is to shark-proof the Internet. That's probably not what they imagined they would be doing when they got their engineering degree.

Coming back to the present, I look at the paused frame one more time. This frame in the Ghibli movie paused in front of me is thoughtfully composed, balancing colour, light and movement to guide the viewer's eye while maintaining a natural, effortless feel. Architecture is drawn with care; character expressions capture the smallest emotional nuances; lighting shifts subtly to indicate mood changes. Nothing feels excessive or unnecessary – each element serves a purpose, creating a sense of harmony where every small detail contributes to the greater whole. Even though I mentioned earlier how Ghibli artists rely on the broad use of light and contrast to convey information, that does not mean that they do not also add extreme levels of detail where required. Individual leaves are painted with the precision of a surgeon, and while the frame looks simple, every detail is well thought out, the complexity helping convey incredibly rich information to our brains.

Perhaps, like with light, there is a duality to everything, from the paintings I spoke about to the animated movie. In some instances, keeping things simple and broad helps us serve the story, while at the same time, other elements may require an incredible amount of detail and precision. Together, these contrasting approaches help us do feats that appear magical, teetering on the impossible.

Here's another example where light is used to do something that feels straight out of a futuristic fiction franchise – something that requires incredible precision, which only light can provide with such ease. There are scientists around me who are using light to control brain cells! Yes, you heard that right. However, this is not for humans, at least not yet. In labs across the world, neuroscientists are using a technique called optogenetics, through which they can use light to literally turn specific neurons on and off like tiny biological switches in animals like mice and flies.

Here's how it works: Every cell in a brain has a complete set of genetic instructions, like a vast library of books.

Scientists have discovered they can add a new 'chapter' to this library – genes from algae that naturally respond to light. When these genes are added to specific neurons, they produce proteins that work like tiny solar panels on the cell's surface.

These light-sensitive proteins are incredibly specific. Blue light causes them to change shape, creating a small opening in the cell's outer layer, like opening a tiny door. When this door opens, ions (electrically charged particles) rush in, causing the neuron to fire. Different proteins respond to different colours of light – some activate with blue light, others with yellow or red. We can even use one colour to turn neurons on and another to turn them off.

But getting light to specific brain cells is tricky; you can't exactly shine a flashlight into the brain. This is where fibre optics come in. We use hair-thin glass fibres to direct light to exact locations in the brain. These fibres work like microscopic light pipes, using the same principles that keep light bouncing along in undersea Internet cables.

The real magic happens when you combine these tools. We can now genetically 'install' light-sensitive proteins in specific types of neurons – for now, just the ones involved in memory, movement or emotion. Then, by delivering precisely timed pulses of coloured light through the fibre optic cables, we can control when these neurons fire.

When you think about it, there's something almost poetic about using light to understand consciousness. Here we are, using the same fundamental force that allows us to see the world to peek into the very mechanics of how we perceive it. It's a beautiful closed loop – light helping us understand how we understand light.

Here's a confession: I find everything I talked about, especially light – the core connecting element that makes all this work – both ridiculous and profound. What do you mean we trap light to send dog memes across the ocean? How can we control entire areas of the brain and even specific neurons by shining light on them? It's like the universe composed a joke so clever and subtle that it took us centuries to get to the punchline.

I'm not alone. One of my favourite authors, Terry Pratchett, had a way of making the ordinary feel ridiculous and profound at the same time. Nowhere is this more evident than in the way he writes about light in his Discworld series, set in a fictional universe which allows him to make fun of Roundworld and also deliver some profound insights.

Light in our world is something swift, intangible, impossible to perceive in motion. But in Discworld, it moves just a little slower – enough that it can be observed, enough that it can be *felt*. It pours like honey over landscapes, *seeps* through the cracks in time and unfurls leisurely across the Disc's sky. This isn't just a clever joke – it's a metaphor that shifts how we see the world. Light, in Pratchett's universe, isn't just a force of nature. It's something *alive*, something with weight and presence, something that takes its time getting where it needs to go.

And because it moves just that little bit slower, it leaves behind a kind of residue, a memory of where it has been. It thickens in the air, pools in corners and gets tangled in the fabric of reality. It allows us to see not just what is there, but what has been there, what lingers just beneath the surface. This is the magic of Pratchett's world. Light isn't just illumination, it's history; it's time itself dragging its feet; it's the world revealing itself one soft glow at a time. And in this, he captures something that's surprisingly real. Because even in our world, light isn't just instant. It bends, it lingers, it filters through dust and mist and it stains the sky long after the sun has set.

Much like in the Ghibli frame in front of me, Pratchett's slow-moving light makes the world feel fuller, deeper, richer. It forces us to notice the spaces *in between*, the moments in which reality blurs just a little, in which magic feels more possible. It allows us to stop and *see*, to watch the way light clings to the edges of things, to realize that a world bathed in slow light is a world where stories have time to settle, to breathe and to unfurl at their own pace. And maybe, just maybe, that's how we're meant to experience it too.

Or maybe it's not really that profound, and we should just make light of it! ☺

9

Creatures of Light and Darkness

The screen goes black. This is not the Blue Screen of Death, thankfully, so it doesn't evoke the same existential dread. It's just a power failure with all the exasperation that evokes.

The Ghibli movie abruptly disappears into darkness mid-frame, taking with it the soft glow that had transformed my living room into a landscape of shadows and highlights. For a moment I sit frozen, my hand still reaching for the chips I'd been about to munch on.

My other senses rush to compensate as my eyes take a break. The whirr of the fan dies, leaving behind a silence that feels almost solid. This is the moment you realize how much background noise your appliances make. Who knew silence could be so *loud*? I also can smell everything much more clearly – including the fact that Oreo needs a bath.

My living room, so familiar just seconds ago, becomes a foreign landscape. I can hear Oreo's tail thumping against the couch; he's completely unfazed by this dramatic shift in our reality. Oreo is a Zen master, calm and meditative until and unless the doorbell rings, whereupon he transforms into a very noisy wannabe killer.

'At least one of us isn't disoriented,' I mutter into the darkness. As if in response, I feel his cold nose touch my hand. He's navigated the pitch-black room perfectly, while I'm still trying to remember if there's anything between me and where I think the candles are.

This plunge into darkness is something that divers, submariners and oil-rig workers who work in the deep ocean face every day. Sunlight fades underwater as light is absorbed layer by layer, creating bands of darkness that stretch from twilight to absolute black. Each layer is a world unto itself, with creatures adapted to that particular level of light, or its absence. At a depth of 200 m or so, there is practically no light, and gradually the ocean turns pitch black. In parts, the oceans are many kilometres deep. But all sorts of marine creatures and marine vegetation adapted to these conditions exist all the way down.

As my eyes are slowly adjusting to the dark, finding shapes in what moments ago was impenetrable blackness, I muse upon the wonders of evolution in the deep sea ...

Deep in the ocean, where sunlight cannot reach, there swims a fish. To our human eyes – if we take it out of its natural habitat – it would be a brilliant crimson. But here's the twist: In the midnight zone of the ocean, this vivid red might as well be an invisibility cloak. At those depths, red light has long since been absorbed by the hundreds of metres of water above, and so there is no red light left to be reflected off and seen by others. You see, water absorbs light selectively – red and orange wavelengths vanish within the first 50 m, while blue and violet light penetrates much deeper. That's because of wavelength – red light has the longest wavelength (and therefore lowest energy) while violet light has the shortest, and short wavelengths penetrate deeper due to the higher energy.

After that brief diversion into camouflage, let's talk about the giant and colossal squids, those legendary leviathans that for centuries were supposed to exist only in sailors' tales of krakens and occasional washed-up remains. Even now, there are large rewards on offer for anybody who can capture a colossal squid alive.

In the darkness of their deep-sea realm, these magnificent creatures have evolved a fascinating way of seeing their world. Their eyes, the largest in the animal kingdom, can grow to the size of dinner plates – not for capturing more light (there is none from outside to capture), but for detecting the faintest bioluminescent signals in the vast darkness. They're searching not for sunlight but for the subtle betrayals of movement: disturbances in bioluminescent organisms that might reveal the presence of prey or predator.

Even more remarkable are the deep-sea seaweeds that have completely abandoned photosynthesis – a rather dramatic career change for a plant! These deep-sea revolutionaries feed on marine snow,[1] a poetic and rather inappropriate name for the constant shower of dead plankton, faecal matter and organic debris that drifts down from the sunlit world above. This disgusting but organically rich mixture of fish poop, dead

---

[1] We probably gave marine snow this name because it looks like snow drifting through dark water and because 'fish poop and dead stuff' doesn't seem quite right – even though that's exactly what it is.

plankton and ocean debris flutters down to the deep sea like underwater confetti. It is crucial for deep ocean life. When resources are scarce, life uses anything that's available to adapt.

Red algae secrete enzymes that decompose this organic matter through a process called extracellular digestion, in which the digestion happens outside their cells. Where their shallow-water cousins transform sunlight into glucose through photosynthesis, these deep-sea innovators have essentially become underwater decomposers, breaking down complex carbohydrates, proteins and lipids into simpler forms that they can use for energy.

Some species have taken this adaptation even further, forming symbiotic relationships with chemosynthetic bacteria – tiny bacteria that act like chemical factories that can synthesize energy from the sulphur compounds seeping from hydrothermal vents found on the ocean floor. These bacteria use a process called chemosynthesis, where they oxidize sulphur compounds to produce energy, much like plants use sunlight in photosynthesis. It's a complete reversal of the typical plant playbook: Instead of reaching up towards the light, these organisms have learned to thrive in its absolute absence, drawing life from the very darkness that would seem to prohibit it.

This hidden world reveals a profound truth about life itself. Where there is an energy source, no matter how alien or obscure, life will find a way to harness it. From the giant and colossal squids to the chemosynthetic bacteria, the dark depths have become a laboratory for evolution's most creative solutions, where the usual rules of land biology are rewritten in the ink of endless night.

It's also a reminder that what we see – or don't see – is just one version of reality. As my eyes continue adjusting to the darkness of my powerless apartment, I notice something peculiar. Oreo, my brave companion who earlier seemed unperturbed by our sudden plunge into darkness, is staring intently at a corner of the room. His night vision, inherited from his wolf ancestors, is far superior to mine, especially in the dark. This may be why he sometimes seems to see ghosts.

Dogs, cats and many such animals have a reflective layer behind their retinas called the *tapetum lucidum* – Latin for 'bright tapestry' – that gives them a second chance at catching any available light. Where I see shadows, he sees a world illuminated by the faintest glow from the streetlights outside.

This makes me think about deer in forests, and how their world looks nothing like what we imagine. Like dogs, deer are dichromats – they see primarily in blue and yellow wavelengths, unlike humans, who see the red wavelengths too. Humans are trichromats, which is actually a rarity in many mammals. It's why it is difficult for your dog to identify the bright red ball in green grass. It may look very contrasting to our eyes, but to theirs, it blends in.

Where we look at a forest and see a kaleidoscope of greens and browns, they see a world of contrasting luminance, where movement matters more than colour. This is why a lion's tawny coat, which seems so obvious to us, blends perfectly into their blue-yellow world. What looks like terrible camouflage to our eyes is perfectly invisible to their (and fortunately for them also the deer's) eyes.

Then there are snakes, who see the world not just in visible light, but in heat. Pit vipers, like rattlesnakes and pythons, possess heat-sensing pits near their eyes, allowing them to detect the infrared radiation emitted by warm-blooded prey. To a snake, a mouse hiding in dry grass is not a shape but a glowing heat signature, a flickering warmth against the cooler landscape. They don't just *see* their prey – they *feel* its presence in a spectrum of light we will never experience.

Meanwhile, back in the ocean, another animal, the mighty mantis shrimp, takes vision to an entirely new level. The mantis shrimp's vision isn't just impressive – it's so excessive it borders on inconceivable. While humans trudge along with a humble three types of colour receptors, this tiny crustacean casually flexes between 12 and 16, granting it access to colours we can't even begin to imagine. It sees UV, infrared and polarized light, detecting details so subtle that, in comparison, what we see seems like a badly compressed image that has lost all detail.

Human scientists eager to understand what reality looks like through its eyes have tried mapping its visual spectrum. The problem? The mantis shrimp sees so much information at once that it doesn't even process colour the way we do; it just recognizes wavelengths instantly, no apparent mental effort required. It doesn't mix red, green and blue like we do. It just *knows*.

The mantis shrimp also has a defensive mechanism that is overkill. It has a punch that isn't just fast –it's *disrespectfully* fast. At nearly 80 km per hour, it moves almost with the speed of a bullet, delivering a strike so forceful that it boils the water around it, creating cavitation bubbles that collapse with a second impact, a burst of heat as hot as the sun and – because this wasn't ridiculous enough – a tiny flash of light. At this point, it's hard not to wonder: who is this shrimp taking notes from? An over-the-top Bollywood action director? An anime fight choreographer? Every strike lands with the timing of a slow-motion hero entrance, leaving science scrambling to recover.

Studying them should be simple. Just put them in a tank, point a camera and record, right?

Wrong.

Because mantis shrimp don't just punch their prey. They punch *everything*. Aquarium walls? Shattered. Research cameras? Demolished. Expensive lab equipment along with the dreams and hopes of the researcher? Gone. Entire experiments have ended with a shrimp fist-fighting science itself and winning. The real research breakthrough here isn't in understanding their punch; it's in figuring out how to observe them before they punch their way out of the study altogether.

At this point, it's worth asking – why? How did this shrimp receive the ultimate VIP pass to the electromagnetic spectrum? Did it strike a deal with the laws of physics? Take notes from a sci-fi cinematographer? Every glance it throws is a full cinematic experience in 32K resolution, while the rest of us are stuck watching the world in greyscale. The difference is so large, human eyes aren't even in the running compared to the mantis shrimp.

I'll give you a moment to let that sink in.

Speaking of sinking in, let's head back into the deeper darker sea again – the kingdom of darkness, where creatures must evolve in a world without sunlight. But some creatures in the pitch-black depths don't just sit around waiting for light – they make their own! It's like a secret hack: Light isn't just something to be absorbed or reflected – it can be *generated*.

That's what they do, and they use it to survive, communicate and deceive in an otherwise lightless world. This ability to produce light biologically – through chemical reactions within their own bodies – feels almost magical, as if these creatures are carrying their own private lanterns through the abyss.

Take the hatchetfish, for instance. It's a small fish that faces a huge danger of being eaten by predators. But this clever little swimmer has basically invented nature's version of stealth technology. Its belly lights up with just enough glow to match whatever faint light is filtering down from above. To predators looking up from below, it's like the fish is wearing an invisibility cloak. Pretty sneaky, right?

And then there's the anglerfish, nature's ultimate catfish. Picture this: You are a little fish swimming in the dark, and you see this beautiful, gentle light bobbing in the distance.[2] 'Oh, how lovely!' you think ... right until you realize that the light is dangling from the forehead of a creature with teeth like a nightmare's nightmare. That glowing lure is actually attached to a fish that looks like it was designed by a horror movie director on a particularly inspired day.

But here's the really cool part. It isn't just a random glow. It's precise biological chemistry that adheres to all the rules

---

2 Also, fun fact that absolutely no one asked for: In some species, the tiny male anglerfish fuses with the female, dissolving all his organs except his reproductive parts, effectively becoming a permanently attached life partner. So, this fish provides the worst relationship metaphor of all time. When the anglerfish was first discovered, the scientists studying it thought it had multiple tumours on its body. It turned out to be the remains of the males that were attached to the female.

of physics – adjusting to how light bends through water, how it scatters, how it illuminates the darkness. These creatures are basically living light shows, performing in nature's darkest theatre.

Thinking of light shows and theatre brings me back to storytelling and animation. It may not look obvious, but bioluminescent creatures *under the sea* actually remind me of one of animation's biggest challenges: How do you make magical glowing hair look real? When Disney tackled Rapunzel's story in the animated movie *Tangled* (2010), they weren't just making another princess movie – they were also trying to solve a mind-bending physics puzzle. How do you make 70 feet of hair both glow *and* move naturally?

You can't just slap a glow effect on some animated hair and call it a day. Real hair clumps when wet, falls with gravity and catches light differently at every angle. Now imagine making that hair not just reflect light but actually produce it. It's like trying to animate a river made of fireflies that also needs to behave like actual hair!

Disney's solution? They created something called Dynamic Wires. Think of it as a virtual hair salon where physics meets magic. They had to simulate 173 different tubes of hair (yes, they counted), which then turned into 140,000 individual strands. That's more individual hairs than most of us have on our heads!

And just like the deep-sea denizens that make their own light, Rapunzel's hair had to illuminate its surroundings in a way that felt natural, even though it was completely magical. When her hair glows, it's not just shining but casting light on everything around it, creating shadows, reflecting off surfaces, all while swishing and swaying like real hair would.

In a way, Disney's animators were doing exactly what evolution did in the deep sea – finding a way to make light work in places where it shouldn't exist at all. Whether it's a bioluminescent fish in the depths of the ocean or a princess with magical glowing hair, both are examples of light apparently breaking the rules, but still somehow following them.

Yet, if there's one thing that never fails to amaze me about us humans, it's our delightful tendency to look at nature's rules as mere suggestions. Like a child who can't resist poking at things they're told not to touch, we've always approached darkness with a 'challenge accepted' attitude. And often, we innovate to find solutions by using our ingenuity, a little bit of luck and science.

I think about this as I fumble for my phone's flashlight, temporarily blinding both myself and an unimpressed Oreo. I also articulate Aparna's Law: The duration of a power outage is directly proportional to how much battery life your phone has left and inversely proportional to how badly you need to check social media. So, the more charge you have, the longer the power cut will last and the more urgent the need to check your Insta, the longer the Wi-Fi will be non-functional.

As the light turns on, and my fear fades away, I muse: Our relationship with darkness has always been a complicated dance of fear and fascination. From the moment early humans first struck stone against stone and sparked fire, we have been driven by an instinctive need to banish the shadows. Those first humans who figured out how to capture fire were not only making light but also rewriting the rules of survival.

Imagine that moment: standing there with a burning branch, watching shadows retreat for the first time, feeling that mix of triumph and wonder that must have made their hearts race.

We didn't stop there, of course. From those first flickering flames, we moved on to oil lamps and candles. I have one of those old-time, antique lanterns sitting on my shelf.

I picked it up from an artist's stall at an exhibition because it seemed essential at the time. It works on kerosene (or diesel at a pinch). You soak the wick and light it after filling the bottom chamber with the oil which soaks up into the wick – it is actually the oil that burns, not the wick. Of course, I don't have any kerosene or diesel handy, so it's purely decorative. If you actually use one, it has a characteristic smell that combines hot metal and burning kerosene.

Looking at it now, I can picture those lantern bearers shown in old Indian horror movies, an old chowkidar walking ahead of travellers through fog so thick you could spread it on toast. Their lights weren't just for seeing – they were tiny islands of certainty in an ocean of unknown, and conveyed to the audience the primal emotion evoked by the single source of light in the darkness.

But it was electricity that really let us thumb our nose at darkness. The light bulb didn't just brighten rooms; it changed how we thought about night itself. Suddenly, darkness became optional, like a dimmer switch for the world. Those first electric lights must have seemed like magic – clean, steady light without smoke or flame, turning night into a sort of second-hand day. No longer were we beholden to the fickle flames of candles or oil lamps. Instead, electric light offered consistency, reliability and the ability to illuminate entire cities at the flick of a switch.

Where once the setting sun signalled the end of productivity, now factories, homes and streets could remain active well into the night. The science fiction writer Nancy Kress wrote a series of books (Beggars in Spain) about a bunch of mutant humans who had only one super-power – they didn't need to sleep and, hence, stay active in the night. So they could be one-third more productive than normal humans. Indeed, by allowing us to work during nights, electric lights roughly doubled our productivity as a species.

The irony doesn't escape me that while I'm sitting here thinking about humanity's conquest of darkness, I'm still slightly startled by every unexpected shadow in my dimly lit room. Perhaps that's why we've never stopped inventing new ways to light up our world – some part of us still remembers when darkness meant danger, when every shadow might hide a predator, when the unknown stretched endlessly into the night.

The shadows in my room remind me that our relationship with darkness – and light – has always been complicated by very human things: ambition, pride and the desire for control. Just as I reach for my phone's reassuring glow when a strange shadow catches my eye, early electricity companies reached for control over how we would illuminate our world. It's a story that seems fitting to think about in this half-light, where shadows and brightness dance across my walls.

Take Thomas Edison and Nikola Tesla – two brilliant minds who saw the same problem through very different lenses. Edison, practical and persistent, approached invention like I approach finding lost things in the dark: methodically searching every corner, testing every possibility. Tesla, on the other hand, built entire machines in his mind first, like a child who can see whole worlds in the shadows on their bedroom wall.

Think about Tesla arriving in America in 1884 with just four cents in his pocket and a head full of revolutionary ideas. It sounds like the beginning of a fairy tale, doesn't it? But what followed was less fairy tale and more Greek tragedy.

Tesla once worked for Edison's company. Edison, already a celebrated inventor, initially welcomed the young engineer. The story goes that Edison offered Tesla $50,000 if he could improve his direct current (DC) generators. Tesla worked tirelessly for months, making numerous improvements to Edison's designs. But when Tesla asked for his payment, Edison dismissed the agreement as a joke, saying, 'Tesla, you don't understand our American humour.' Instead, he offered a modest raise of $10 per week above Tesla's weekly salary of $18.

This perceived betrayal marked the beginning of their deteriorating relationship. Tesla resigned immediately and, for a time, found himself digging ditches to survive. But his belief in the superiority of alternating current never wavered. While Edison was focused on improving existing technology, Tesla was reimagining the entire system from the ground up.

Instead of the DC system that Edison relied on, Tesla created an alternating current (AC) system. While our laptops and smartphones work off batteries that use DC, our household and industrial power supply is usually AC. Alternating current is more efficient and requires less maintenance.

Imagine, for a moment, water flowing through pipes. Direct current is like a steady stream flowing in one direction – simple, straightforward, but losing strength the further it travels, like a river gradually slowing as it moves through flat land. Each little loss of power along the way adds up, which is why Edison had to build power plants every few city blocks. It's a bit like having to place water pumping stations every mile or so because the water pressure keeps dropping.

Direct current works well enough for low-voltage devices like smartphones, but AC? Alternating current is like a dance, with electrons swaying back and forth in a rapid rhythm – 50 or 60 times every second. At first, this might seem like an odd way to deliver power. Why make electricity do the cha-cha when it could just flow straight ahead? But here's the magic: this dancing current can do something DC can't – it can transform.

Think of it like this: Alternating current can be 'stepped up' to extremely high voltages for long-distance travel (where losses are minimized), then 'stepped down' again for safe use in our homes. It's like being able to turn a gentle stream into a powerful jet for travelling long distances, then transforming it back into a gentle flow when it reaches your tap. This is why Tesla's system could power entire cities from a single plant, while Edison needed a power plant in practically every neighbourhood.

Nature itself uses something like AC – brief, powerful surges of electricity – to send massive amounts of energy through the air. Tesla grokked this natural wisdom and built an entire power system around it. Meanwhile, Edison, brilliant as he was, stuck to his DC guns like someone insisting that cars should still be pulled by horses – not because it was better but because he'd invested in a lot of stables.

I glance at Oreo, who cares nothing for the electrical currents flowing around us, and smile inwardly as I muse on how something so complex has become so mundane that even its gentle hum is part of our daily lullaby ...

Today, as we're surrounded by devices powered through AC, Edison's fear tactics to push DC during that war of currents seem almost comically dramatic. He staged public demonstrations electrocuting animals to prove AC's dangers, turning light – that thing we'd sought for safety since our cave-dwelling days – into a weapon of fear. It's like using a nightlight to scare children instead of comforting them.

But here's the thing about truth: like light itself, it has a way of penetrating even the deepest shadows. Tesla's AC system simply worked better.

Edison's DC power plants had a simple but fatal flaw – they couldn't send electricity very far. After travelling about 1.5–2 km, the power would weaken too much to be useful. This meant that if a city wanted electricity, it couldn't rely on just one power plant; it needed dozens, each one supplying small areas. The bigger the city, the worse the problem. Streets would be lined with power stations, burning coal, taking up space and driving up costs. It was a system that worked, but only if you didn't think too big. As cities grew, it became clear: This wasn't the future of electricity.

While Edison's DC plants needed to be as common as corner stores, Tesla's alternating current could leap across vast distances, carrying power from a single source to thousands of homes. It was like comparing a collection of candles to a single, brilliant sun.

The 1893 World's Columbian Exposition in Chicago must have been something to behold. Imagine stepping into a fairground blazing with electric light, all powered by Tesla's AC system. In an age when many people had never seen an electric light, it must have felt like walking into the future. Even now, when I flip a light switch without thinking, I can feel an echo of that wonder.

Yet triumph and tragedy often dance together. I told you in the beginning that this story was not a fairy tale but a Greek tragedy. Tesla, the brilliant, unstable Tesla, tore up a contract worth millions to save industrialist George Westinghouse and ensure his system would succeed.

The story behind this sacrifice still makes me pause: Westinghouse had agreed to pay Tesla a royalty of $2.50 for each horsepower of electrical capacity sold. It doesn't sound like much, but with electricity spreading across America like wildfire, these royalties would have made Tesla one of the wealthiest men in the world. But then came the financial panic of 1893, sending the economy into a tailspin. As I think about this, my phone buzzes with a news alert about current market fluctuations, reminding me how some patterns in history never really change. J.P. Morgan's banks, backing Edison's interests, gave Tesla and Westinghouse an impossible choice: Lose the company, or renegotiate the contract. Here's where the story turns from business to something more profound. Tesla's story showcases how some people chase glory while others chase dreams.

Upon hearing that his royalties might bankrupt the only company willing to champion his AC system, Tesla tore up his contract, surrendering his personal fortune so that his vision of electricity for everyone could survive. It wasn't about money – it was about lighting up the world.

The tragedy is theatrical beyond reality: The man who gave us the power system we still use today died nearly penniless, while the man who fought against this superior system, Edison, made a fortune and kept it. Capitalism favours profitable businessmen over genius but broke scientists.

Edison's own company had to bow to the inevitable march of progress. Sitting here in my AC-powered home, I think about how General Electric's (GE's) story mirrors our own struggles with change –sometimes we have to let go of what we thought was right to move forward. General Electric, despite Edison's fierce resistance to AC power, found itself at a crossroads, and it had a stark choice: Adapt or become obsolete.

General Electric chose adaptation. In 1892 it merged with Thomson-Houston, a company that had already embraced AC technology. It's almost like a plot twist in a business drama – Edison's company needed the very technology he had fought against so viciously to survive and grow. The newly formed GE didn't just quietly adopt AC; it dove headfirst into it, developing transformers, generators and transmission equipment for the AC system that Edison had once called dangerous and impractical.

The irony deepens when you realize that GE's growth into a global powerhouse was built largely on AC power. All those dramatic demonstrations with electrocuted animals, all the publicity campaigns against AC's dangers, all of Edison's stubborn resistance – in the end, his own company's survival depended on embracing Tesla's vision. The very company that continued to carry a trace of Edison became one of the primary vehicles for his rival Tesla's dream of universal electrical power. Sometimes, I suppose, our legacies take paths we never could have imagined, flowing like electricity itself along unexpected circuits.

There's a bittersweet lesson here about the difference between being right and being rewarded. Tesla chose his dream over his fortune, wanting his light to reach everyone, even if it meant personal darkness. Meanwhile, the business empire that grew from Edison's resistance to progress thrived by embracing that very progress.

Sitting in the darkness, I think of how humans, the same species that can have moments of pure brilliance, also hold the capacity for so much darkness. Edison was a brilliant innovator, a visionary who brought electric light to the world; yet he was also a man who waged one of the most cut-throat smear campaigns in scientific history.

It's a paradox that feels deeply human. The same mind that illuminated entire cities also cast its own dark shadows. The same species that crafts symphonies, maps the cosmos and writes poetry is also capable of cruelty, destruction and self-sabotage.

Ancient cultures understood this long before psychology put a name to it. Light and dark exist in all of us. Sometimes, we charge ahead, driven by discovery, by wonder, by the desire to make the world better. Other times, we hold on too tightly, driven by fear, ego or the refusal to let go of control. Circumstances and our unique stories bring out the deeper parts of our own self – sometimes coming as a surprise even to ourselves.

Nowhere is this tension more evident than in our relationship with darkness itself.

I think about how our ancestors must have felt when night fell. For them, darkness wasn't just the absence of light, but the realm of uncertainty and the unknown, filled with predators and perils they couldn't see. Each innovation, from the first controlled fires to modern LED bulbs, pushed those shadows (and the predators within them) a little further away.

But here's the catch – as we've gotten better at pushing back the darkness, we're starting to realize we might have pushed too far. The stars I used to count on summer nights are fading behind the haze of city lights due to light pollution. I remember travelling back from the mountains of Bhutan once. We could see the stars and constellations clearly from the little villages nestled quietly in the Himalayan valleys. As we approached India with its cities again, the stars disappeared as the towns around us were lit up with artificial lights.

Nocturnal creatures – those creatures that are adapted to living in the dark – who have spent millions of years evolving to thrive in darkness, find their world increasingly invaded by our artificial day. Even our own bodies are rebelling; here I am, watching 'just one more episode', while my brain protests against the blue light keeping it artificially alert.

Perhaps it's time we learnt another lesson: Despite all our efforts to banish the dark, it never truly disappears. We chase certainty, yet uncertainty persists. And perhaps it is better that way. Still, there's something deep inside us that prevents us from letting go of our fear of the dark.

Even right now, there's something about a power outage

that makes me temporarily revert to living like our cave-dwelling ancestors, stumbling around and stubbing my toes on furniture that's been in the same place for years.

This momentary disorientation takes my mind to yet another thought. It's fascinating how quickly I've become helpless without my – and most humans' – primary sense: sight. It makes me think about those who navigate the world without visual cues, partially or fully. How does the world reveal itself to them? How different is their experience from those of us who are fully sighted?

I think about my recent trip to the metro station, where I recently noticed something I'd seen but never truly understood before. You may have seen it too – a network of yellow tactile paths that wind through the platforms like rivers of raised dots and lines. Curious about it, I researched what it was and was impressed with what I discovered.

The formal name for this system is Tactile Ground Surface Indicators (TGSI), but that term hardly captures how these simple patterns have transformed urban navigation for millions.

It is a strikingly simple yet elegant system. Unlike boards and signs written in words that we can see and read, this system speaks in a language of texture and contrast. The yellow paths I mentioned earlier consist of two main patterns: straight, parallel lines called 'guide paths' that mean 'go forward', and raised dots arranged in a grid pattern called 'warning blocks' that signal 'stop or change direction'. The bright yellow isn't just for aesthetics – it's like nature's warning colours, serving those with partial vision. It reminds me of how the deep-sea creatures we talked about earlier use bioluminescence as both signal and warning.

This system converts the entire metro station into essentially a massive tactile map. If implemented properly, these paths create reliable routes through what might otherwise be overwhelmingly complex spaces. They lead to commonly frequented spaces like ticket counters, where Braille markings on automated machines allow independent ticket purchase. They are also present to guide users to escalators and elevators, each marked with easy-to-decipher touch-based tactile signs. When we reach the platform edges, there are additional textured warnings with a different pattern distinct from the directional markers to alert users that they're approaching the track.

The brilliance of the system reveals itself further in the details. At every intersection where paths cross, the pattern changes – creating what I imagine feels like a paragraph break in this tactile story, a moment to pause and consider your next direction. Near platform edges, the warning tiles are deeper – about 600 mm compared to the usual 300 mm – providing an extra margin of safety. It's like a friend's hand on your shoulder urging caution, the deeper tiles giving you twice the usual space to register the approaching track. Even the spacing of the raised elements is carefully calculated: close enough to be detected by a cane sweep, but far enough apart to prevent wheels (whether from luggage, wheelchairs or strollers) from getting stuck.

The whole system speaks in multiple dialects simultaneously – through texture for those reading with their feet or canes, and through colour contrast for those with partial vision, without creating obstacles for anyone else.

In addition to the tactile paths, the metro also uses yet another tool – sound. In metro stations, there are regular audio announcements in multiple languages – Hindi, English and regional languages depending on the city. These announcements don't just call out station names; they provide real-time information about train arrivals, platform changes and even which side the doors will open. Some stations have also begun implementing audio-tactile maps at entrances – raised diagrams that speak when touched, providing an audio overview of the station layout.

The Indian system reflects lessons learned from earlier implementations worldwide. Japan pioneered tactile ground indicators in 1967 when Seiichi Miyake invented them for a crossing near a school for the blind in Okayama. The first installations were simple dots, and the more sophisticated line-and-dot system evolved through years of user feedback. When Indian metros began adopting these systems, they could build on decades of international experience while adapting to local needs.

At the heart of these systems is not just accessibility but also safety, which means engineering these paths requires surprising precision. The height of the raised elements must be enough to be detectable – typically 5 mm – but not so high as to create a tripping hazard. The materials must be durable enough to withstand millions of footsteps yet maintain their tactile properties. They need to be slip-resistant even in monsoon conditions. Even the installation process is crucial – any gaps or misalignments can disrupt the path's readability.

Here's what strikes me. These aids provide something more than accessibility. They provide independence, allowing someone who's born with or has a visual disability the ability to navigate a public space in a safer manner.

Perhaps what's most remarkable is how these features have become seamlessly woven into the daily fabric of metro life. Watch any busy station during rush hour, and you'll see how the tactile paths serve everyone. Elderly passengers often use them for extra stability, their canes finding reassurance in the regular patterns. Young children treat them like balance beams, turning their commute into an impromptu game. Even rushed office workers, phones in hand, unconsciously follow these paths through familiar stations. What was designed as an accessibility feature for the visually challenged now allows all passengers to intuitively navigate these spaces.

The paths also create unexpected moments of community. Regular commuters often notice when someone seems to be hovering uncertainly near these paths and may offer assistance without being asked. During peak hours, you may see people naturally forming lines along the guide paths, using them as informal queuing markers. The paths have become a shared reference point in the organized chaos of urban transit.

I'm not saying this system is perfect; no system is. Some stations have gaps in coverage, especially in older sections. Maintenance can be inconsistent, with worn patterns or damaged tiles sometimes taking time to be replaced. It is also a reminder that we need to do more and improve constantly.

But what warms my heart is that the underlying principle – that everyone deserves to navigate public spaces with dignity and independence – has literally become the ground of our urban infrastructure, one yellow tile at a time.

This integration of different ways of experiencing the

world isn't new. In fact, the story of TGSIs is incomplete without another, older story. Let me tell you a story of innovation that opened the world of education, music to people who couldn't see alphabets and writing: the story of Braille. It begins, surprisingly, on moonless battlefields where soldiers crouched in shadows, where even a match struck to read an order could mean death.

We need to go to the night-time battlefields of the Napoleonic Wars. While fighting in the dark, a simple flash of light to read an order could let the enemy know your position and draw enemy fire.

Enter Charles Barbier. An artillery captain, Charles Barbier understood intimately how the inability to communicate in darkness could cost lives. His solution was brilliant in its simplicity: if soldiers couldn't safely use light to read, why not let them read with their fingers? His 'night writing' system used raised dots pressed into paper, like tiny constellations you could feel rather than see. This solution, which came to be called *écriture nocturne* or 'night writing', was revolutionary for its time.

Barbier's system used a grid of 12 dots (two columns of six) to represent different phonetic sounds in French rather than individual letters. Each character was a specific combination of these dots, creating what we might now call a tactile phonetic alphabet. The system was complex by design – it included symbols for common military commands and contained around 195 different symbols total. The idea was sound: if soldiers could read messages in complete darkness, they could coordinate movements without revealing their positions to the enemy.

The execution wasn't quite there yet though. The 12-dot grid was too large for a single fingertip to encompass, requiring readers to feel around each character to understand it – imagine trying to identify a constellation by touching the stars one at a time. The phonetic nature of the system also made it difficult to use for those unfamiliar with formal French pronunciation, and its military-specific symbols had limited use outside combat situations.

But sometimes imperfect innovations find their way to exactly the right person. Barbier saw potential for his system beyond military applications, and so, in 1821, he demonstrated his system at the Royal Institute for Blind Youth in Paris.

Among the students that day was a curious 12-year-old named Louis Braille, who had lost his sight at age three in a terrible accident in his father's workshop. The accident occurred when young Louis was playing with an awl – a sharp leatherworking tool – that slipped and struck his eye. The resulting infection spread to both eyes, leaving him completely blind.

Louis was a brilliant student with an analytical mind. He was also fond of music, with his tactile sense enriched from a young age through learning musical instruments. Perhaps that's why when Barbier demonstrated his system, Braille immediately grasped both its potential and its limitations. Over the next three years, through methodical experimentation and refinement, he transformed Barbier's military code into something more elegant and with far greater utility.

The modification was both tiny and huge. Braille's key insight was reducing the grid size to six instead of

12. Here's why this mattered: The human fingertip, when stationary, could reliably distinguish six dots in a rectangular configuration – three rows of two dots each.

This solved one of the biggest problems of the previous system. A smaller matrix meant that readers could feel entire characters at once rather than having to explore them gradually.

Next came the complexity. Since Barbier's system was designed for military purposes, it was too complex for widespread use, especially due to its reliance on using phonetic sounds. Braille realized that focusing on individual letters rather than phonetic sounds would make the system more intuitive and universal.

Braille's invention is elegant in its simplicity – six dots, like six stars in a constellation, arranged to create an entire universe of meaning. Each dot could be either raised or flat, and in this binary dance of presence and absence, he found 63 different ways to speak. It's rather like nature's own coding system, DNA, where four simple bases combine to write the story of life. The mathematical beauty of his system reminds me of long-time chess players who can play multiple games simultaneously without seeing any of the boards they are playing on.

By 1824, at just 15 years old (the same age many of us were struggling with basic algebra), Braille had mapped out a system where the first 10 letters used only the top four dots, and then the next 10 included the bottom left dot, like verses building in a poem. The last six letters completed the pattern with the bottom right dot – a progression so logical it feels almost musical. Perhaps that isn't surprising, given he played both the cello and the organ; he understood that patterns, whether in music or mathematics, speak to something fundamental in our minds.

For Louis Braille, this wasn't just about reading letters – it was about opening every door that blindness had closed. He continued refining the system until he published the final version in 1837. He created notations for mathematics and music, refusing to accept that blindness should limit anyone's access to knowledge or art. Imagine being able to read music with your fingertips, to feel a symphony encoded in dots before playing it on your cello.

The Greek tragedy that struck Tesla also struck Braille. His adversary wasn't Edison or another scientist, but the system of teaching itself. Despite its clear advantages, Braille's system faced significant resistance, particularly from sighted teachers who found it difficult to learn and preferred teaching methods they could easily read themselves. Think about it, those who could see were deciding how the blind should read! It would take until 1854, two years after Braille's death, for his system to be officially adopted by the Royal Institute, and several more decades for it to gain widespread international acceptance. It is perhaps a reminder that sometimes our greatest gifts to the world only bloom after we're gone.

It seems that history loves geniuses; it just doesn't like paying them. Before we start to admonish Braille's and Tesla's contemporaries for their stupidity and cruelty, we must acknowledge that it's not that future generations have a better moral compass and superior intelligence. We are benefiting from the power of hindsight. Even if Tesla or Braille lived today, their inventions would probably not get venture capital funding because of being too ambitious and impractical.

When this system came to India, there were even more unique challenges for widespread use, particularly given the complexity of Indian writing systems. Initially, different Indian languages had different ways of representation in braille. However, in 1951, we standardized Bharati Braille, a single Braille script which represented the diverse Indian languages. Like Louis Braille's original work, it was an elegant solution to an intricate challenge, accommodating multiple distinct writing systems – from Devanagari to Tamil, from Bengali to Malayalam – under one unified script.

The breakthrough of Bharati Braille was that it maintained a consistent structure across languages while respecting the unique phonetics of each script. Unlike English Braille, which compresses common words and sounds into contractions for efficiency, Bharati Braille took a different approach: no contractions, no shortcuts – each letter was represented distinctly.

This decision wasn't just about simplicity; it was about accessibility. Indian languages already had a layered phonetic structure, and adding contractions would have made learning even harder. Instead, Bharati Braille offered a direct representation of each character, making it easier to teach and learn, no matter the language.

Another key challenge was how to handle vowel modifiers, a major feature of Indian scripts. In many Indian languages, the sound of a consonant changes based on attached vowel signs. In print, these signs might appear before, after, above or below the consonant – a design that didn't translate well into Braille's linear structure. Bharati Braille tackled this problem by assigning separate Braille symbols to vowels and vowel markers, allowing blind readers to interpret them in a clear, logical sequence.

The solution was so elegant that someone who learns Bharati Braille for one Indian language can read content in other Indian languages, even if they don't understand what they're reading, just as the sighted can read a text written in a familiar script even if it's in an unfamiliar language.

Today, Braille's six-dot system remains largely unchanged from his original design – a testament to its refined efficiency. While modern technology has created new ways for visually impaired people to access information, Braille continues to be vital for true literacy, allowing direct reading and writing rather than just listening. The system has been adapted for virtually every language on Earth, proving the universality of Braille's insight about how humans can read by touch.

As the world is moving to be far more connected digitally, it's a useful reminder. How we interact with the world isn't just about what we see – it's about how information is structured, how spaces are designed and how experiences are shaped to be understood in different ways.

I remember thinking about these different ways of experiencing the world during digital accessibility workshops. These experiences were enlightening, helping me learn how differently people interact with the same websites and apps that I use daily. What looks like a simple image to me might be a barrier to information for someone using a screen reader if it lacks proper alternative text. A sleek, minimalist button that I find aesthetically pleasing might be nearly impossible to use for someone navigating with a keyboard instead of a mouse. The same webpage exists simultaneously as a visual design, an audio experience, a set of keyboard commands and a structured document, and it needs to work coherently in all these forms. When we forget this, we unknowingly create barriers in spaces that should be open to everyone.

Even the smallest choices have an effect on accessibility. For me, this hit home when I was writing my first scientific paper. I'd chosen what I thought were aesthetically pleasing red and green colour combinations for my graphs, until I read an article that pointed out something I'd never considered: my carefully chosen colours would be indistinguishable for someone with red-green colour-blindness.

It was a wake-up call. Here I was, trying to communicate scientific findings to the world, while unknowingly making my graphs unreadable for roughly 8 per cent of male readers. The solution, of course, was to use patterns and colours that work for the colour-blind.

I'd chosen colours that looked striking to my eyes without considering how others might perceive them. The figures existed in my personal visual reality but failed in their fundamental purpose – to communicate information clearly to all readers. I wasn't alone in making this mistake.[3]

Thankfully, in my case, I had realized the problem long

---

[3] Many scientific discoveries have been buried because of unclear explanations or boring graphs. Then, of course, there is also the famous story that suggests that Space Shuttle Challenger crashed due to a bad graph. The engineers had diagnosed that a cold start on a morning with low temperatures would create blowback problems and they had graphed it, but it was done very badly and it was ignored. So it seems like a problem affecting many eminent people before me.

before those figures were used. The solution wasn't difficult – just using different colour combinations and adding patterns or shapes to differentiate data points. But it made me realize how many decisions we make based on our own way of perceiving the world, without considering other perspectives. In science, where accurate communication of information is paramount, this oversight could exclude certain readers from access to important discoveries.

These realizations ripple out like circles in a pond. Once you start noticing how spaces and information can be perceived differently, you see opportunities for more inclusive design everywhere – from the colours in scientific figures to the layout of websites, from navigation cues in public spaces to the way we describe images in digital documents.

Sometimes these revelations come in unexpected moments, like the time I went to watch a 3D movie with my friends from my college, many of whom wear glasses. As I casually picked up my 3D glasses, excited about the film, I watched one of my friends awkwardly trying to fit them over her regular prescription glasses. It was clearly uncomfortable, and she spent the first few minutes of the movie adjusting them, trying to find a position that wouldn't give her a headache. When the movie ended, I was left with the experience of watching a satisfying movie, but she was left with a lingering headache. It took this experience for me to understand why she didn't like watching 3D movies.

I felt a sudden jolt of realization – here was something I'd never had to think about because it had never been my problem to solve. I'd been to dozens of 3D movies and had never once considered what that experience might be like for someone who needs prescription glasses. The cinema offered standard 3D glasses, wheelchair accessibility, closed captions for some screenings – but the simple challenge of wearing 3D glasses over regular glasses remained an awkward afterthought.

This is how privilege sometimes works – not through malice or deliberate exclusion, but through the simple fact that we don't notice what we don't have to deal with. Like my earlier obliviousness to colour-blind-friendly scientific images, or my easy navigation of poorly designed websites, these blind spots in our awareness shape the world we build. The things that don't affect us personally become invisible, not because we don't care, but because we've never had to care.

It reminds me of that old saying about how fish don't know they're in water. When something works seamlessly for us, it's easy to assume it works that way for everyone. But the world is full of invisible barriers that some people have to navigate every day while others pass by unawares – from the height of countertops to the timing of crosswalk signals, from the format of digital documents to the design of movie theatre accessories.

But recognizing these patterns isn't a story of failure; it's a catalyst for innovation. Every time we notice a gap, an awkward design or a space that doesn't work for everyone, we identify an opportunity to make things better. When we listen to people describe their experiences, when we pay attention to how different bodies and minds navigate the same spaces, we expand our understanding of what's possible.

This awareness has driven some of our most useful innovations. Curb cuts – those gentle slopes at street corners

– were designed for wheelchair users but now help parents with strollers, travellers with luggage, delivery workers with carts and countless others. Text-to-speech technology, originally developed for visually impaired users, now helps millions of people listen to articles while commuting or cooking. When we start looking at the world this way, every limitation becomes an invitation to innovate, every constraint an opportunity to create something better.

Even the vibrating alerts on our phones, initially an accessibility feature for deaf users, have transformed how all of us navigate social spaces. My phone is permanently on vibrate mode, as I'm mortified by the thought of it ringing in public under any circumstances.

These solutions, born from paying attention to diverse needs, don't just solve problems for some users but make the experience better for everyone. They remind us that accessibility isn't about creating separate, special accommodations – it's about recognizing that human experience comes in many forms, and good design should embrace that diversity.

The power flickers back on. A brief hum fills the apartment as forgotten machines whir back to life. The fan starts spinning again, its steady rhythm rejoining the world of the audible. The TV screen glows, the Ghibli movie resumes mid-frame, washing the room in warm, familiar colours. Shadows shift, bending and reshaping under the sudden return of artificial light.

Oreo blinks, stretching as if nothing had happened, as if the darkness had been a mere pause, an intermission rather than an intrusion. He turns his attention back to the screen, unconcerned, unbothered.

For a brief moment, when the world was swallowed in black, I had seen something – not with my eyes, but with something deeper. A clarity that didn't come from brightness, but from absence. From stillness. From the way the mind reaches for understanding when there are no distractions left to hold it back.

In the glow of the screen, the room feels smaller again. Defined. Contained. But just minutes ago, in the void of light, it had expanded into something vast, something unknown – an ocean of shadow, an uncharted territory. I think of the creatures that thrive in places where no light reaches, who have evolved to see in ways we cannot fathom: The deep-sea fish cloaked in red invisibility, the mantis shrimp deciphering a spectrum beyond my imagination, the snake tracing the heat of something hidden. I think of Rapunzel's glowing hair, an impossible light source made real through both science and story. Of Braille's six-dot alphabet, a language of touch that reveals a world beyond sight. Of Tesla and Edison, locked in a battle where light wasn't just invention but control. Of the stories we tell, the ones that illuminate and the ones that distort, shaping our perception like light bending through water.

Maybe we don't really see the world. Maybe we just see what the light allows us to. This struck me as a profound realization, until one day when I walked full speed into a glass door and cursed not only myself, but the shopkeeper, the cleaner, the inventor of glass and the first person who ever thought transparency was a good idea.

I glance at Oreo, still watching the movie, his tail flicking lazily. Maybe he understands something I don't. Maybe he always has.

Some things are only visible when the lights go out.

And maybe – just maybe – that's when we see the most clearly.[4]

---

[4] Like the fact that my apartment is way messier and more cluttered than I had thought.

10

Eat, Love and Do Science

The electricity is back, and suddenly the day is drawing to a close. This book started with breakfast, and now it's time to organize dinner. Almost unnoticed, the whole day has passed, while I've been busy doing the things I love and telling stories I love. This brings me both joy as well as a small twinge of sadness.

When I'm doing the things I love, time always speeds up. Sundays in particular seem to pass quickly, leaving behind memories like the aftertaste of a favourite dessert. Perhaps the beauty of the day also lies in its ephemeral nature.

I am absent-mindedly making dinner when I'm shaken out of my existential musings about Sundays by a sharp pain in my finger. I realize that I've accidentally cut myself. Oreo points his impossibly human eyes in my direction and gives me his patented '*again?*' look. I never thought that I would be living with a Michelin food critic dressed in a furry black and white coat complete with tail. If this was a cooking show, I can imagine Oreo saying: 'Well, she's got talent, but she's too clumsy. I expected better, honestly.'

I shrug and quickly wash my hand, making sure that I wash the wounded area gently but thoroughly. After drying my finger, I instinctively reach for the ever so useful Band-Aid to place over my cut. I peel it open and stick it on, and I am suddenly transported back to my childhood.

Being a very accident-prone child, I remember my mom calling me 'Rana Sanga' after the famous warrior whose body was covered with over 80 wounds. He often went into a new battle while still recovering from previous injuries. I was given that nickname[1] not because of my battle skills – which were zero – but because I kept getting new wounds before the previous ones had healed. This is also why my memories of childhood include my limbs looking like a mosaic of multiple Band-Aids.

As I fumble with this one, I marvel at this tiny miracle. It looks so ordinary now, a fixture in every home, purse and first-aid kit. But ordinary things sometimes have extraordinary stories, if only we know how to look in the right places.

The story of how the Band-Aid came into being is one for the storybooks and worth telling. It's a tale of love, ingenuity and kitchen chaos.

But first, let me indulge in a little rant.

One of my pet peeves is a popular misconception that many people have. People often think that scientists, or people who invent things, are emotionless. If anything, scientists –

---

[1] The word 'nickname' by the way comes from the old English 'ekename', which meant 'also called as'. Over time, the word changed to 'nickname', as it is used now. Did you need to know this? Well, now you can always whip this factoid out as an icebreaker.

at least some of them – are deeply romantic. It is extremely funny that they are often portrayed in the media as high-IQ robots with no heart.

Science, much like poetry and art, is one way of writing love letters to the universe.

In reality, it takes a special kind of romanticism to stand at the very edge of human knowledge and step forward with hope into the unknown, instead of despairing about the vastness of our ignorance.

Inventors and scientists look at the universe, and love it enough to find new things. That is the emotion that drives a lot of science. Overall, science and scientists get a really bad rap for no reason.

Scientists are often among the most optimistic members of sapiens. We (if I may include myself in that category) look at something that hasn't happened, or something we don't understand, and we say 'hasn't happened *yet*' and 'we don't understand it *yet*'.

Many of us spend our entire working lives trying to make that 'something' happen; quite often, we succeed, and even when we fail, we learn a lot from our failures.

Okay, rant over. Back to the Band-Aid.

The invention of the Band-Aid was a gift of love. Band-Aids feel like a modern invention, but in reality they were invented in 1920, even before we had discovered antibiotics. (Many ancient cultures like India did use turmeric, which is a natural antibiotic to treat cuts, but the mechanism of why this worked wasn't clear).

So, imagine if you will a kitchen, resembling mine in terms of the chaos. Josephine Knight Dickson is cooking food.

Like me, Josephine was prone to accidents, often cutting and/or burning herself in the kitchen. But remember, this is 1920; antibiotics are unknown and Band-Aids don't exist. This means that any cut or nick has the potential to become infectious and potentially life-threatening. (Antibiotic resistance and the rise of superbugs may take us back to that era, however, and the thought sends shudders down my spine.)

The closest thing to a solution to this problem at that time was to tie a strip of cloth around the wound and hope it held together.

Enter Earle Dickson.

Unable to stand by and watch his wife suffering, Earle Dickson, took it upon himself to build something. Earle Dickson was a tinkerer, and also – and this is important – a cotton buyer for Johnson & Johnson.

Being a tinkerer, he decided to take matters into his own hands and develop a solution for reducing his wife's pain and risk of infection. The problem: Open wounds are likely to cause complications and simply tying strips of cloth over a wound isn't reliable or efficient. Medical tape and gauze existed, but these are cumbersome to use. Materials like cotton tend to stick to the wound painfully, and cloth fibres can carry infections.

His idea was simple. He took a strip of surgical adhesive tape, laid a piece of antiseptic cotton gauze down the centre and covered it with crinoline to keep the layers from sticking together. The result was a roll of ready-to-use adhesive bandage.

Just like that, something that would become a staple in every first-aid kit around the world was invented.

He proudly gave this roll to Josephine, who could cut it to the desired lengths, and use it to neatly cover her cuts. No more relying on strips of cloth.

The story didn't end there. The invention worked so well for Josephine that Earle decided to share it with his boss at Johnson & Johnson. They saw its potential and began producing small test batches of what they called BAND-AID® adhesive bandages.

At first, sales were slow. The concept was so new that people had to be taught how to use the product. But Johnson & Johnson believed in the product and stuck with it. They sent door-to-door salesmen to demonstrate the bandages to doctors and pharmacists, who gradually began introducing them to their patients and customers.

By 1921, BAND-AID® bandages had officially hit the market. In the early years, they were handmade and sold in rolls, just as Earle had invented them. But in 1924, Johnson & Johnson developed machines to mass-produce pre-cut, individually wrapped bandages – recognizably the version we use today.

So that's the incredible story of an invention that started as an act of love in Josephine's kitchen and quickly became a household essential. Earle Dickson's invention became so successful that he was promoted to vice president at Johnson & Johnson and eventually joined the company's board of directors. Decades later, in recognition of his contribution to improving everyday life, Earle was inducted into America's National Inventors Hall of Fame.

While I have been telling you the story of loving someone enough to be inspired to the point of invention, I am 'cooking' that classic comfort food – instant noodles, and my noodles have been quietly simmering away. A simple miracle in a pot.

At night, when I'm too tired to cook anything elaborate, I tend to reach for instant noodles. The PhD student in me resurfaces. Instant noodles have helped me survive late-night research breakdowns, existential crises at 2.00 a.m., entire chapters of my thesis being written (and deleted) and more silly caffeine-fuelled decisions than I care to admit.

It's funny how something so basic, and so reliable, was once considered a luxury.

Let me set the scene.

It's post-war Japan. Food is scarce, and people line up for hours just to get a bowl of fresh ramen.

A Japanese businessman, Momofuku Ando, watched this struggle and wondered: 'Why should a warm meal be so difficult to get?'

While warm ramen bowls were traditional, making them fresh and warm for a large population was expensive, not to mention cumbersome. One of the biggest issues was that noodles could not be stored easily, and had to be freshly made by hand right before being boiled. If the noodles were made and stored earlier, they tended to either become too soft or too dry, and would not absorb moisture correctly. Driven by his own love for ramen, and also by love for his shattered, defeated country, Ando was determined to change this.

He spent many months experimenting. He steamed, dried, boiled and tried combinations of all of these. He kept trying, and he kept failing. The resulting product either turned to mush or stubbornly remained uncooked. Then, one evening, as he watched his wife frying tempura, inspiration struck–what if he flash-fried the noodles?

Ando quickly tried it, and something incredible happened. The hot oil created tiny holes in the noodles, turning them into sponges. Later, when hot water was poured over them, they reabsorbed moisture instantly, bouncing back to life like they had never been dried in the first place.

And drumroll, instant ramen was born!

It was cheap, easy to make and, most importantly, it worked. No more waiting in long lines for a warm meal. Just boil water, add noodles and pretend life is under control.

But Ando didn't just invent a product.

He believed food was more than just a means for survival – for him, it was an act of love. A warm meal is more than just sustenance; it is comfort, it is care, it is proof that someone thought about you before you even sat down to eat.

Oreo looks at me, as if to remind me to make sure that I don't burn the noodles. I don't think he would be happy if I had yet another accident. I quickly turn the stove off, watching as the flames flicker and die beneath my pot of noodles. A few minutes later, as I look for a fork, I realize that I have overlooked yet another everyday miracle, an invention that is so ubiquitous that it's found everywhere from student dorms to space shuttles – instant fire.

Fire. Perhaps one of the most powerful and earliest of human inventions.

Today, I can get fire on command, control the flame and do so without rubbing two stones together until I get a spark. However, for most of human history, fire wasn't just something you turned on – it was something you fought to keep.

It was a stolen secret. A crime. A rebellion against the gods.

as never meant to be ours.

s stolen and given to humanity; or so many of our aim. Prometheus, Māui – different names, same story. erpret those myths as people's way of acknowledging ver of fire – something so powerful that it had to be ne origin.

e wasn't just an invention of survival; it was an invention e and love. When early humans first tamed fire, it didn't st keep them warm or ward off predators – it changed everything. It softened food, making it easier to chew and digest, which meant that those who lost their teeth – whether from age, injury or illness – could still eat and survive. Long before modern medicine, before writing, before civilization itself, people were already looking out for one another. The simple act of cooking meant that food could be shared with the weak, the young and the old.

At the same time, the myths also tell us how difficult it is to control and keep a fire alive.

Because fire, if left alone, dies.

A slight wind could either make the fire die or make it so uncontrollable that it destroyed everything. It's as if the gods themselves didn't want us to have that power.

The fire we were never meant to have became the one thing we could not live without – our control of fire turned humans into the species that controls everything on the planet.

So, for us humans, the real miracle wasn't in stealing fire. It was keeping the fire alive and under control.

Because – as I have said before – fire, if left alone, dies.

And so we fed it. We learnt how to control it. We built hearths. We gathered around it. We used it to cook food, not just for ourselves but for others.

Fire gave us light in the darkness; it gave us a way to off predators and to cook food in ways that allowed u absorb nutrients at a faster rate.

Fire did something else too.

It gave us one of our oldest love stories.

Dogs.

Before we domesticated cows, before we planted crops, before we even built the first villages, we had already welcomed dogs into our hearts. They were the first animals we domesticated, not because we set out to, but because love, in its simplest form, finds a way to do mysterious things.

We fed them, and they guarded us. We cared for them, and they stayed. It wasn't a single moment but a slow dance that began around ancient fires. Archaeological evidence tells us that somewhere between 15,000 and 40,000 years ago, when humans were still hunter-gatherers following herds across vast landscapes, wolves began to shadow our camps. Not large, aggressive wolves that posed a threat, but the more curious ones, perhaps the ones who were a little less fearful, a little more willing to venture closer to human settlements.

They might have started as scavengers, drawn to the refuse of human camps. But something remarkable happened. Instead of driving them away, our ancestors began to share their fires, their food and, eventually, their lives with these wild creatures. It wasn't a calculated decision to create a new domesticated species. It was simpler than that – more instinctual, more emotional.

We fed them, and they guarded us. We cared for them, and they stayed.

We just. *Loved* them.

And love, as it turns out, is the greatest evolutionary advantage.

The story of dogs and humans is written not just in our hearts, but in DNA. Recent genetic studies tell us that dogs split from wolves somewhere between 15,000 and 40,000 years ago, though the exact timeline remains debated. This wasn't a single event, but rather a long dance of coevolution, happening across different regions as humans and wolves began their tentative partnership.

Early humans didn't just select for strength or speed; we started selecting for something else, something much softer: companionship. This selection is visible in what scientists call 'paedomorphosis' – the retention of juvenile features into adulthood.

The wolves that stayed with humans began to change physically – their snouts became shorter, their teeth smaller, their ears more floppy. These 'cute' features (that are found in modern dogs but largely absent in wolves) are actually the physical manifestations of selecting for friendlier, more socially tolerant animals.

And before you say it, no, this isn't just me projecting my love for dogs onto all of humanity.

Archaeological evidence shows this transformation wasn't just physical. The earliest dog burials tell us that these animals were already seen as more than just tools. They were buried with care, sometimes with grave goods, suggesting they held a special place in human society. The evidence of this ancient partnership is written in bones found across the world. In a cave in Germany, archaeologists discovered a 14,200-year-old burial where a person was laid to rest with their hand resting on a young dog. In Israel, a 12,000-year-old burial revealed an elderly person cradling a puppy. These weren't just practical arrangements – they were expressions of attachment, of bonds that transcended species.

This relationship also reshaped dog brains. Dogs developed an unprecedented ability to read human emotions and gestures – they're better at interpreting human pointing than even chimpanzees, our closest relatives. Other domesticated animals have similar abilities, so perhaps this has something to do with time spent interacting with humans. Domestic cats, for instance, have vocalizations they only use to communicate with humans, not with other felines. Dog brains release oxytocin – often called the 'love hormone'[2] – when they make eye contact with their human companions, just as human brains do when we look at our loved ones, including our dogs.

Over generations, wolves became dogs, and dogs became family. We bred them to be guardians, workers and herders, but also to be companions. They went from creatures that lurked at the fire's edge to those curled up at our feet. Studies show that dogs have developed specialized muscles in their faces that wolves don't have – muscles that help them make those puppy-dog eyes that we find so irresistible. It's a physical adaptation specifically for communication with humans.

And maybe, at the heart of it all, that's what truly changed the world – not just fire, not just survival, but the simple act of choosing to care for another being. We can see this

---

[2] This is a simplistic and not quite accurate description, but it'll have to do.

ancient bond reflected in cultures worldwide. Dogs are faithful companions in so many cultures, appearing for example in Norse and Irish mythologies as well as the Mahabharata. The relationship between humans and dogs appears in the earliest written records, suggesting it was already ancient history when writing began.

Every puppy that wags its tail at the sight of its human, every dog that offers comfort with a gentle nuzzle and every faithful companion that watches over a sleeping child is carrying forward this ancient legacy. They are the inheritors of a bond that has helped shape human history, a partnership that began before written records, before cities, before all the trappings of civilization.

It was the beginning of something profound – not just domestication, but a kind of love that would help define what it means to be human.

Many thousands of years later, Oreo is watching me and using all of these abilities to channel his disgust at my egregious crime of breaking his favourite treat into two. But he also knows that he can do this primarily because of one fundamental truth: He is loved.

As he sits there, his head tilted just so, his eyes wide and bright with the same curiosity that must have driven his ancestors toward those first fires. He doesn't know that his kind once helped humans shape civilization itself. He only knows that he is loved.

And sometimes, that is enough to make a difference.

I saw this first-hand with Oreo. When he first came to me, he was a tiny puppy terrified of the urban world around him. Traffic sounds, which to his sensitive ears must have seemed like constant thunder, would send him into a panic. It makes perfect sense – dogs can hear frequencies up to 65,000 Hz, compared to our measly 20,000 Hz. What seems merely loud to us can be overwhelmingly intense for them.

But just like those ancient wolves who gradually learnt to trust humans, Oreo needed a way to safely experience this scary new world. So I became his sanctuary, carrying him in my arms during our walks. It was a mutual adaptation. For me, it was an unexpected weight training regimen; for him, it was a safe vantage point from which to process the overwhelming sensory input of city life. In my arms, he could observe the world while knowing he was protected.

Slowly, step by step, day by day, his trust grew along with his confidence. The sounds that once terrified him became background noise. The streets that seemed threatening became exciting paths to explore. Now, walks are his greatest joy – tail wagging, ears perked, eager to discover what new adventures await us around each corner.

Looking at Oreo now, confidently trotting down the same streets that once terrified him, my heart fills with a joy I can't quite explain.

And the truth is – he saved my life.

I remember the small puppy who was once too scared of everyone deciding, against all his fears, that it was safe to crawl into my lap and fall asleep. The tiny, traumatized child who clung to me, because traffic sounds terrified him, but he *knew* he was safe when he was hugging me. My big boy, who nudges me gently toward the kitchen if I've forgotten to eat for too long.

He saves my life every single day.

And I'm so grateful to be loved by him[3].

He is a reminder that sometimes, we all need a little help, and it's okay to let someone carry you when the world is extremely scary outside. That's what love does – it makes us find ways to help, to protect, to nurture. It drives us to solve problems we might otherwise dismiss as unsolvable.

Perhaps this is one of humanity's most remarkable traits – our capacity to care for the vulnerable. We look at those who need help, whether they're scared puppies, fragile babies or struggling adults, and we think, 'How can I make this better?'

I laugh when I hear modern-day podcast hosts shouting about 'survival of the fittest' as if it justifies callousness or cruelty. They're missing something fundamental about human evolution. Our species didn't become 'fit' by abandoning the vulnerable – we thrived precisely because we learnt to care for them. We survived not by being the strongest, or the fastest, but by being the ones who stayed behind to help others catch up, inventing ways to do that more efficiently along the way. We were the ones who carried our injured, who nursed our sick, who protected our young. The ones who looked at vulnerability and saw not weakness, but an opportunity to make things better.

Love, you see, has a way of driving innovation that nothing else quite matches. When we care deeply enough, we find solutions that might otherwise have remained undiscovered. Sometimes, these solutions change not just our own lives but the lives of countless others.

Consider the story of Alexandre Lion and his baby

---

3 Big emotions.

incubators. In late nineteenth-century France, premature babies were considered 'weaklings' with little chance of survival. Most doctors had given up on them. But Lion, driven by a profound sense of care and responsibility, looked at these tiny lives and refused to accept their fate.

He drew inspiration from an unlikely source: chicken incubators. Poultry farmers had been successfully using warming devices to hatch eggs for years. Lion wondered: 'If we could keep chicks alive this way, why not human babies?' He developed sophisticated incubators with temperature controls, air filtration systems and humidity regulation.

But the medical establishment wasn't interested. Hospitals wouldn't fund what they saw as a futile endeavour. So Lion did something extraordinary: he put his incubators on display at fairs and exhibitions, charging admission to see the babies being cared for. The admission fees funded the care of premature infants whose mothers couldn't afford treatment.

These 'child hatchery' exhibits might seem shocking to us today, but they saved hundreds of lives. They showed the world that premature babies could survive with proper care. Each tiny life saved was a testament to what love – even love for strangers – could accomplish.

Just as Oreo's need for safety led me to carry him until he found his courage, Lion's compassion for these fragile lives led him to create something revolutionary. Both stories remind us that sometimes, the greatest innovations come not from the pursuit of profit or fame, but from the simple, powerful desire to care for another living being.

The modern neonatal intensive care unit (NICU) bears little resemblance to Lion's exhibition incubators, but both share the same core purpose: protecting vulnerable lives until they're strong enough to face the world on their own. And, in the end, isn't that what love does best?

Sometimes, that love means protecting not just vulnerable lives, but vulnerable ideas – the ones that challenge what everyone 'knows' to be true. Ideas that, like premature babies in the nineteenth century, might be dismissed as too weak to survive in the harsh world of established science.

This brings me to Vera Rubin, who looked at the stars and saw something nobody else wanted to see. In the 1970s, when she began studying the rotation of galaxies, she noticed something that shouldn't be possible. The outer edges of galaxies were spinning as fast as their centres. According to all known laws of physics, this would mean that these galaxies should have been tearing themselves apart. But they weren't.

Rubin refused to accept the conventional wisdom that her observations must be wrong. Instead of dismissing what she saw, she protected and nurtured this anomaly, this vulnerable idea that challenged everything astronomers thought they knew about the universe.

Her persistence led to one of the most profound hypotheses in science: She postulated the existence of dark matter, an invisible substance that could make up more than 85 per cent of all the matter in our universe.

But the path wasn't easy. As one of the few women in astronomy at the time, Rubin faced constant scepticism. When she first presented her findings, many male colleagues suggested she must have made a mistake. Some refused to believe her data entirely. After all, how could a woman who had been barred from Princeton's graduate astronomy programme (which didn't accept women until 1975) possibly discover something that had eluded the field's most respected minds?

Yet Rubin persisted, carefully documenting her observations, gathering more evidence, studying galaxy after galaxy. She didn't just protect her own vulnerable ideas but became a fierce advocate for other women in science. She demanded that institutions include women on their committees and in their speaker lists. She mentored young female astronomers, creating safe spaces where their ideas could grow strong enough to survive in a field that often seemed designed to exclude them.

Her work not only revolutionized our understanding of the universe but also demonstrated why protecting vulnerable voices matters. Just as Lion showed that 'weakling' babies could thrive with proper care, Rubin proved that ideas dismissed by the establishment might contain profound truths about our universe. Her discovery wasn't just a triumph of observation – it was a testament to the power of nurturing what others might discard.

Today, dark matter remains one of science's greatest mysteries, invisible yet crucial, holding our galaxies together. And perhaps that's a fitting metaphor for all the quiet acts of support and protection that hold our human world together – invisible forces of love and care that keep everything from flying apart.

Speaking of challenging what the establishment thinks is possible, I have to tell you about Mangalyaan, India's Mars Orbiter Mission that made the global space community sit up and take notice.

In 2013, when the Indian Space Research Organisation (ISRO) announced its plan to send a spacecraft to Mars, many dismissed it as overly ambitious. After all, more than half of all Mars missions had failed. Even NASA, with billions in funding, had lost spacecraft to the red planet. And here was ISRO, planning to do it on a budget that was smaller than the production cost of the Hollywood movie *Gravity*.

The entire mission cost $74 million, a fraction of NASA's MAVEN Mars mission that cost $671 million. The comparison is striking: ISRO was attempting interplanetary travel in what NASA spent on its catering budget.

But like Vera Rubin studying her 'impossible' galaxy rotations, ISRO's scientists didn't let conventional wisdom deter them. They found innovative ways to do more with less. Instead of building entirely new technologies, they adapted and refined their existing satellite technology.

The team developed what they called 'frugal engineering'. Every gram of weight had to earn its place. When they couldn't afford specialized space-rated components, they found ways to adapt and 'ruggedize' commercially available parts. The spacecraft's autonomous systems were designed to be both robust and resource-efficient, capable of making split-second decisions with minimal computing power.

What's truly remarkable is that they didn't just succeed with such tight constraints – these constraints drove innovation. The solutions they developed were more than just workarounds; they were elegant approaches that often worked better than their more expensive alternatives.

One of their most ingenious innovations was in the mission's trajectory design. Instead of following the typical straight path to Mars using powerful rockets and lots of fuel, the team used a technique they called 'phasing orbits'. Think of it like climbing a spiral staircase instead of taking an elevator. The spacecraft gradually expanded its orbit around Earth, gaining energy and altitude in stages.

Starting in Earth orbit, the spacecraft executed a series of carefully timed engine burns to stretch its orbit into increasingly elongated ellipses. Each manoeuvre was precisely calculated to use minimal fuel while maximizing the gravitational assist from Earth. It was like a cosmic dance where every step had to be perfect; a single miscalculation could have sent the spacecraft off course.

The liquid apogee motor (LAM) engine was another testament to frugal innovation. This engine had to restart after sitting dormant in space for 300 days – something it wasn't originally designed to do. Rather than designing a new engine (which would have been prohibitively expensive), the team developed a series of innovative heating algorithms and backup plans to ensure the sleeping engine would fire when needed.

The on-board autonomy system was equally clever. With communications delays between Earth and Mars ranging from four to 24 minutes, the spacecraft needed to make critical decisions on its own. The team developed a system that could perform complex orbital calculations, adjust the spacecraft's orientation and manage power systems with minimal computing resources. They wrote extremely efficient code that could run on radiation-hardened but relatively basic processors.

For power management, they developed what they called

'spatial solar power management' – a way to angle the solar panels that maximized power generation while minimizing mechanical stress on the system. During eclipse periods, when Mars blocked the sun, the spacecraft had to operate on stored power while keeping all critical systems running. The power management algorithms they developed were so efficient that they ended up extending the mission's life far beyond its planned duration.

And on 24 September 2014, Mangalyaan made history. India became the first Asian nation to reach Mars, and the first nation in the world to do so on its maiden attempt. The spacecraft slipped perfectly into Mars' orbit, ready to begin its study of the red planet's atmosphere.

Another remarkable aspect of Mangalyaan was that it didn't just succeed – it kept on succeeding. Originally designed to last six months, it continued functioning for eight years, far exceeding its planned mission life. Like those premature babies in Lion's incubators who grew up to live full lives, Mangalyaan proved that with the right care and innovation, even the most ambitious dreams can survive and thrive.

The mission didn't just demonstrate India's space capabilities; it showed the world that groundbreaking science doesn't always require earth-shattering budgets. Sometimes, the most remarkable achievements come from working within constraints, from finding creative solutions when the obvious path is too expensive or impossible.

One of the most powerful images from Mangalyaan's success was the photograph that went viral after the spacecraft reached Mars: a group of Indian women in colourful saris, sporting bindis, hugging each other and celebrating the success of the mission in ISRO's control room. The image shattered multiple stereotypes at once – about women in science, about Indian capabilities, about what a 'typical' space scientist looks like.

Women scientists led crucial aspects of the Mangalyaan mission. Ritu Karidhal, known as the Rocket Woman of India, was the Deputy Operations Director. Nandini Harinath worked on mission design and helped navigate the spacecraft to Mars. Anuradha T.K., then the senior-most woman officer at ISRO, oversaw the satellite communications. Minal Sampat led the team that built the project's software systems.

These women weren't just participating – they were leading, innovating and problem-solving at every level.

Their success highlighted something profound about inclusivity in science: when you remove artificial barriers and let talent flourish, remarkable things happen. These women didn't succeed because they tried to fit into a pre-existing mould of what a space scientist should be – they succeeded because they were given the opportunity to be themselves while pursuing excellence.

In every one of these stories, these impossible inventions and discoveries, I can see the same thread – love taking different forms. Not just the romantic kind, but the kind that makes humans look at an impossible problem and think, 'There has to be a way.'

For the young engineers building rockets in makeshift facilities, it's love for the pure challenge of it, the kind that keeps you up at night wrestling with equations because you can't let go of a problem. They're not doing it for wealth or fame; many left cushy software jobs to work in tiny startups where success is far from guaranteed. But there's something about the problem that pulls them in, makes them want to solve it.

This is how love drives innovation – not always dramatically or obviously, but in the quiet persistence of people who can't let go of a problem, who keep pushing boundaries not because they have to, but because something in them needs to know what's possible.

This is not to say that any of it is ever easy.

Today, it's easy for me to tell these stories as if they were inevitable – as if Lion knew his incubators would save thousands of lives, as if Vera Rubin knew her 'impossible' observations would reveal the existence of dark matter, as if the ISRO team knew Mangalyaan would not only reach Mars but function for eight years.

But that's not how it happened at all.

When Lion was setting up his incubator shows, most of the medical establishment thought he was wasting his time – and possibly being cruel – by giving parents false hope. Every morning, he must have walked into those exhibitions carrying not just the responsibility of tiny lives, but also the weight of doubt. What if everyone else was right? What if he was wrong? But he kept going, because the alternative was doing nothing.

When Vera Rubin saw galaxy rotations that defied known physics, she didn't know she was going to discover dark matter. She just knew that something wasn't adding up, and she couldn't pretend otherwise – even when colleagues suggested she must be wrong, even when her findings were dismissed. She kept measuring, kept calculating, insisting that what she saw was real, with no guarantee that she'd ever be proven right.

The ISRO team worked with budgets so tight that a single mistake could have ended everything. They couldn't know if their innovative solutions would work in the harsh environment of space. Every decision carried the weight of possible failure – not just personal failure, but national disappointment. When they used Earth's gravity to slingshot their spacecraft toward Mars, they couldn't have been absolutely certain it would work. They had calculations, yes, but in space exploration, even the smallest oversight can be catastrophic.

What keeps people going in the face of such uncertainty? It's not confidence in success but something deeper. It's the love that says 'this matters enough to try', even when you don't know if you'll succeed. It's the hope that makes you show up every day to work on something that might fail, because the possibility of making a difference is worth the risk of disappointment.

We need to remember this when we face our own uncertainties. The people we now call visionaries or pioneers weren't people who could see the future – they were people who kept going when they couldn't see it at all. They loved what they were doing enough to face the possibility of failure, the weight of doubt, the voice of criticism that said they were wasting their time.

That's what real innovation looks like – not a clean line from idea to success but a messy path of uncertainty, lit only by the stubborn hope that something better is possible. And sometimes, that's enough.

Yes, we must talk about what needs to improve. The gaps in

our systems, the failures of our institutions, the places where we fall short – conversations about these are vital. They drive progress, push for accountability and help us identify where change is needed most urgently.

But there's another story that deserves telling, one that often gets lost in the louder narratives of crisis and criticism. It's the story of people who wake up every morning and choose to work on problems they might never solve completely. In labs across India, scientists are working on sustainable energy solutions, not because they'll single-handedly solve climate change but because every small improvement matters. In government schools, teachers are finding innovative ways to make science accessible to their students, not because they'll revolutionize education overnight but because each child who understands something new is a victory worth pursuing.

These people aren't the ones with the dramatic success stories that make headlines. They're the quietly persistent researchers who spend years studying a single protein that might someday help us understand a disease better. They're the software developers who work to make their applications more accessible for users with disabilities, knowing they might never achieve perfect accessibility but believing that each improvement matters. They're the environmental scientists collecting data year after year, building the long-term understanding we need to protect our ecosystems.

Their work isn't flashy. It doesn't promise instant solutions or revolutionary breakthroughs. Instead, it offers something else: the steady accumulation of knowledge, the gradual improvement of systems, the patient pursuit of better possibilities. They're not working toward a moment of dramatic triumph; they're committed to the long, slow work of making things better, step by small step.

In research labs, classrooms, offices and fields across the country, people are choosing to tackle problems that are too big for any one person to solve. They're choosing to make their small contributions to a larger solution they might never see completed. There's something profound in that choice – in knowing that the work is bigger than you are, and doing it anyway.

This too is love – the kind that shows up day after day, that celebrates small victories while acknowledging how far there still is to go, that chooses to keep working even when the finish line isn't in sight.

And perhaps somewhere there is a girl who tells stories of science – not because she's the best scientist, or the most accomplished writer, but because she loves the way discovery and wonder weave together into tales worth telling. She sits with her dog, trying to capture the magic of innovation and persistence, of minds that refused to accept 'impossible' as an answer.

In case you haven't caught on yet (and if you haven't, I worry about your detective skills), that's me. Yes, the same person who just spent several chapters comparing everything from wireless networks to baby incubators to my dog's anxiety issues. The one who somehow managed to connect the entire history of dog domestication to my puppy's fear of traffic sounds.

I write these stories between explaining to Oreo why he can't eat my dinner (he remains unconvinced), and trying to convince myself that I'll actually cook instead of ordering takeout (I remain unconvinced). Sometimes I get it wrong. Sometimes I fall into the very traps I'm trying to write about – making neat stories out of messy realities, or viewing breakthroughs through the distorting lens of hindsight.

But I keep writing anyway. Not because I'm sure these stories will make the world better, or change anyone's mind, or accomplish anything grand at all. I write them simply because I love telling them. There's something about the way science and stories intertwine that makes me want to share them, even if I'm not sure where they'll lead or who might read them.

Like those scientists working on problems they might never solve, I'm just doing what I love – telling stories about discovery, about persistence, about people who kept going even when they weren't sure where they'd end up. And maybe that's enough.

Even if Oreo thinks I should spend more time throwing his ball instead.

So now, sitting on a blanket with my noodles, staring up at the stars, let me be philosophical once more.

The amount of love hidden around us every day is so much more than what can be conveyed in the stories I told you. I told you about Josephine's Band-Aids and Ando's noodles, about Mangalyaan's slingshot into the orbit of Mars and Lion's incubators for babies. About the first wolves who chose our fires over their forests, starting a 40,000-year-long dance of co-evolution that gave us the dogs we know today.

But we didn't explore all the hidden acts of care that hold our world together – like how a mother's body silently selects thousands of proteins to feed her baby exactly what it needs while keeping harmful molecules away, or how scientists spend decades studying a single protein fold, knowing they might never see the lives their work saves.

We didn't talk about the researchers who developed mRNA vaccines decades before we needed them, or about the teachers who buy extra pencils with their own money. Or the nurses who hold patients' hands when their families can't be there, or the people who translate medical documents in their spare time so others can understand their diagnoses.

There are countless people who show up every day to push the boundaries of what's possible, not because they have to, but because they care enough to try. Every corner of our world holds these invisible forces that keep things from flying apart. Some were crafted by evolution over millions of years, some emerge from late nights in labs and some arise from quiet moments of kindness that no one ever sees.

As for me, as I look at this little boy curled up on my leg, looking up at me as if I am his whole world, I feel like I can take a little bit of the love he has for me and wrap it around me like a warm blanket.

And I think to myself ...[4] At the core of it, love itself is very much like dark matter. I can't explain it, and yet I know it's real because in the absence of it, the universe itself doesn't make sense.

Good night.

---

[4] What a wonderful world.

Epilogue: The Gaps between Certainties

There are many legends about kings who have placed impossible questions before sages and court jesters, only to receive answers that radically shifted the king's perspective. The common element in all these stories is that you've probably missed the real point if you think you understand the question in its entirety.

It's the same when you're doing science or writing about it: The moment you think you understand something completely, you've probably missed something important.

In this book, I've tried to capture the wonder I feel when I look at the world, from the way light dances through water to how love drives innovation. I've attempted to show how ordinary everyday things, ranging from the bandage on my finger to the wireless signals carrying these words in digital form, can hold extraordinary stories within them.

I've done my best to be diligent.

My browser history is a graveyard of research paper PDFs, each a rabbit hole leading to a dozen others. For every story that made it onto these pages, a hundred more live in my notebooks, scribbled between coffee stains and hastily drawn diagrams. Some are tales of accidental discoveries made while looking for something else entirely. Others are ideas that seemed brilliant at midnight and questionable by morning. There are experiments that failed spectacularly, only to lead somewhere unexpected.

For example, we learnt to see inside atoms by studying duck eggs. And that, famously, a melted chocolate bar led to the invention of the microwave – though that was perhaps less about scientific insight and more about someone trying to hide evidence of snacking in the lab.

Sometimes I felt like I was trying to map a city at night with only a flashlight. I could see either where I was or where I was going, but never both at once. Each story illuminated a small patch of ground, but I'm acutely aware of all the streets I never explored, the windows I didn't peer through and the conversations I didn't overhear.

There are probably cracks in the narratives too. In tracing the delicate threads connecting one discovery to another, I may have pulled some too tight, or missed others entirely. Each chapter is like a snapshot of a moving target, sharp in places, blurred in others, already slightly outdated the instant it's captured.

So if you find mistakes, and I'm sure some of you will, please tell me. Nothing would make me happier than learning where I went wrong, because that means I get to learn something new. That's the most exciting part of science – every mistake is an invitation to see the world with a little more clarity.

There is a Japanese art form called kintsugi, which entails repairing broken pottery with gold – but the cracks aren't hidden. Rather, they're celebrated as part of the object's history. I'd ask for your grace to do something analogous.

Consider these chapters as invitations to wonder, rather than definitive accounts. Use them as starting points for your own explorations, your own questions, your own discoveries. If something seems wrong or incomplete or oversimplified, follow that thread. See where it leads you. After all, the beauty of science isn't in having all the answers, it's in having new questions to ask.

I've tried to be conscious of my own biases. But I'm certain they've crept in anyway, colouring which stories I chose to tell and how I chose to tell them. Perhaps I've unconsciously favoured explanations that appeal to my aesthetic sense, or overlooked evidence that didn't fit my preferred narrative. Science is done by humans, after all, and written about by humans too – complete objectivity may be something we strive for but can never fully achieve.

I'm hardly immune to the seductive pull of a good story. Like everyone else, I carry my own constellation of biases, my own favoured ways of seeing the world. When I find an explanation I like, I tend to hold onto it a bit too tightly, seeing confirmatory evidence everywhere while gently pushing contradictions to the margins. It's a deeply human tendency. As a species, we're natural storytellers, pattern-seekers, meaning-makers. We want things to make sense, to fit into neat narratives with clear causes and effects.

However, the universe doesn't care what story we find most appealing or which hypothesis we've grown attached to. Sometimes the most beautiful theories crumble in the face of a single, stubborn fact. Sometimes the explanation we've championed turns out to be quite wrong, and the perspective we dismissed holds the key to understanding. If there are narratives in this book that fall apart because of new evidence, I would still be happy that I tried.

Science, at its heart, is just structured play. It's about asking 'what if?' and then trying to find out. What if we looked at this differently? What if we tried again? What if this thing we think we understand actually works some other way entirely?

These chapters are verbal experiments. They are attempts to capture lightning in a bottle, to translate the wonder of discovery into words. Some may have worked better than others. I've probably oversimplified some things, overcomplicated others and drawn conclusions that might not stand up to peer review. But isn't that how science works? We try things out, see what happens, learn from what goes wrong and try different things.

So I'll keep telling stories, making mistakes, learning and sharing what I learn. If there's one thing writing this book has taught me, it's that knowledge, like light, becomes more interesting when we realize how little we see.

Looking back, I'm struck not by what I know, but by how much more there is to learn. Each fact we uncover is like a new star becoming visible, revealing not just itself but the vast darkness around it where other stars must be waiting.

Let's see where the next set of questions takes us.

Acknowledgements

Writing your first book, I've discovered, is rather like trying to build a spaceship while simultaneously mugging up rocket science and attempting to overcome a fear of heights. Most days, I felt like that student who hasn't studied enough for exams but still hopes for a miracle. The fact that you're holding this book means I somehow managed to pass – though there were plenty of moments when that seemed highly doubtful.

I've been extraordinarily lucky to find myself surrounded by people who not only read what I write with genuine enthusiasm (even those early drafts that made about as much sense as my first attempts at cooking) but who also maintain that perfect balance between 'you're doing great!' and 'what nonsense is this?' – a combination that every writer needs but not everyone is fortunate enough to find.

I am deeply grateful to Chiki Sarkar and the team at Juggernaut for believing in this book and bringing it to life. I want to express my heartfelt thanks to Devangshu Datta, whose editorial guidance has been invaluable in shaping this book. Your insights and expertise helped transform my ideas into their clearest, most compelling form. Thank you for helping me turn this dream into reality.

At the heart of this support system are my parents, Gayatri Agarwal and Mahesh Chand Agarwal, and my brother Vikram Singh Pathania, who nurtured my curiosity from an early age with their stories about the world around us. They've perfected the art of looking interested when I go on and on about science at the dinner table, and their encouragement and confidence in me have been the foundation upon which this book was built. I know they're already planning to tell everyone, 'See, my daughter/sister is not just a doctor, she's gone and written a book also!' – and honestly, I wouldn't have it any other way. I am blessed with a large, lively family whose energy and joy have been a constant source of inspiration and comfort. Their love, laughter and endless conversations have enriched my life and, by extension, these pages.

I carry with me the profound influence of my late uncle Mr Vishvanathan Nair, who constantly debated with me without ever making me feel small. I hope his legacy lives on in these pages.

This blend of wonder and systematic thinking was further nurtured by the remarkable teachers I've had from school all the way to my PhD. Dr Sanjay Sane, my non-advisor whose advice I often sought, and Prof. K. Vijayraghavan, whose lectures never felt like lectures and who taught me how well science can be communicated by passionate storytelling.

Sourabha, thank you for the push I needed to begin writing this book. Avantika, thank you for always pushing me to make up more things (a dangerous suggestion to make to a scientist). Your ability to get excited about science rivals my own, which I didn't think was possible.

Chandni, thank you for your genuine enthusiasm for this

book from its earliest conception and connecting me to my editor, which literally made this whole thing possible.

To my improv friends and theatre companions – thank you for teaching me how to weave narratives that connect with people. To Shweta and Laxmi, your fresh perspectives have been invaluable, even when you had to tell me, 'Maybe simplify that a bit more?' And to Vineet, who paradoxically keeps pushing me to write more while taking his sweet time to read what I send – your encouragement (and eventual feedback, whenever it arrives) means the world. I'm pretty sure you'll finish reading this acknowledgment sometime next year.

To Umesh, my first port of call for all things science, thank you for continuously enriching my scientific world and being an ever-ready sounding board – even if our 'quick chats' about science somehow always turn into three-hour debates. To the other Aparna, my voice of reason, thank you for your honesty and for saying what needs to be said. Both Siddharths – Bharath and Kankaria – thank you for being my constant champions, believing in me especially during times when I forget to believe in myself.

Pooja, thank you for keeping me grounded when my thoughts drift too far into the clouds. Krupa, who graciously gave me her laptop when mine broke down and has been extremely enthusiastic about every update I give her about the book, and Deepten, who does the same and also continues to give the best hugs always.

To Pritha, Rohit, Jyothi – some of my oldest friends – thank you for proving that true friendship knows neither time nor distance. Though adult life means we meet but once a year, your presence in my life remains constant and cherished.

Our reunions might be rare, but they're worth every minute of scheduling Tetris it takes to make them happen.

And Arun – my most enthusiastic (and possibly least objective) reader, who has convinced himself that every single thing I write is brilliant. Thank you for being there through it all. I'm still not sure whether to trust your reviews, but I'm eternally grateful for them. Your support is something I wish every writer gets to experience.

To Oreo, my faithful companion and wonderful co-conspirator – thank you for being my guiding force through this journey. You may not understand science but you understand the importance of regular snack breaks, and sometimes that's more valuable.

And now, to those whom I've inevitably forgotten to mention (because writers can be terribly forgetful, and I claim this rightfully) – if you're reading this and wondering why your name isn't here … well, overconfidence in my memory has led many astray before you. Please know that my gratitude runs deeper than my memory. Feel free to remind me of this oversight at the next social gathering – I promise to look appropriately embarrassed, dramatically slap my forehead in recognition and make it up to you with coffee, conversation and possibly baked goods.

Writing can be a solitary endeavour, but I have never felt alone thanks to all of you. I hope you like this book!